Information and Instructions

This shop manual contains several sections each covering a specific group of wheel type tractors. The Tab Index on the preceding page can be used to locate the section pertaining to each group of tractors. Each section contains the necessary specifications and the brief but terse procedural data needed by a mechanic when repairing a tractor on which he has had no previous actual experience.

Within each section, the material is arranged in a systematic order beginning with an index which is followed immediately by a Table of Condensed Service Specifications. These specifications include dimensions, fits, clearances and timing instructions. Next in order of arrangement is the procedures paragraphs.

In the procedures paragraphs, the order of presentation starts with the front axle system and steering and proceeding toward the rear axle. The last paragraphs are devoted to the power take-off and power lift systems. Interspersed where needed are additional tabular specifications pertaining to wear limits, torquing, etc.

HOW TO USE THE INDEX

Suppose you want to know the procedure for R&R (remove and reinstall) of the engine camshaft. Your first step is to look in the index under the main heading of ENGINE until you find the entry "Camshaft." Now read to the right where under the column covering the tractor you are repairing, you will find a number which indicates the beginning paragraph pertaining to the camshaft. To locate this wanted paragraph in the manual, turn the pages until the running index appearing on the top outside corner of each page contains the number you are seeking. In this paragraph you will find the information concerning the removal of the camshaft.

More information available at Clymer.com
Phone: 805-498-6703

Haynes Publishing Group
Sparkford Nr Yeovil
Somerset BA22 7JJ England

Haynes North America, Inc
859 Lawrence Drive
Newbury Park
California 91320 USA

ISBN-10: 0-87288-527-5
ISBN-13: 978-0-87288-527-1

Cover art by Sean Keenan

SHOP MANUAL

MASSEY-FERGUSON

MODELS
MF255 — MF265 — MF270 — MF275 — MF290

Tractor serial number is stamped on a name plate attached to the instrument console.
Engine serial number is stamped on the side of engine cylinder block.

INDEX (By Starting Paragraph)

DUAL DIMENSIONS

This service manual provides specifications in both Metric (SI) and U.S. Customary systems of measurement. The first specification is given in the measuring system perceived by us to be the preferred system when servicing a particular component, while the second specification (given in parenthesis) is the converted measurement. For instance, a specification of ''0.28mm (0.011 inch)'' would indicate that we feel the preferred measurement is the metric system of measurement and the U.S. Customary equivalent of 0.28 mm is 0.011 inch.

CONDENSED SERVICE DATA

	MF255 Gasoline	MF255 Diesel	MF255 Diesel	MF265 Gasoline
GENERAL				
Engine Make	Perkins	Perkins	Perkins	Perkins
Engine Model	AG4.212	AD4.203	A4.236	AG4.236
Number of Cylinders	4	4	4	4
Bore	98.4 mm (3.875 in.)	91.5 mm (3.6 in.)	98.4 mm (3.875 in.)	98.4 mm (3.875 in)
Stroke	114.3 mm (4.5 in.)	127 mm (5.0 in.)	127 mm (5.0 in.)	127 mm (5.0 in.)
Displacement	3.48 L (212 cu. in.)	3.33 L (203 cu. in.)	3.87 L (236 cu. in.)	3.87 L (236 cu. in.)
Electrical System	12 Volt, Negative Ground			
Forward Speeds	8 or 12			
Reverse Speeds	2 or 4			
TUNE-UP				
Firing Order	1-3-4-2	1-3-4-2	1-3-4-2	1-3-4-2
Valve Clearance, Cold				
Intake	....	0.30 mm (0.012 in.)	0.30 mm (0.012 in.)	....
Exhaust	....	0.30 mm (0.012 in.)	0.30 mm (0.012 in.)	....
Valve Clearance, Hot				
Intake	0.30 mm (0.012 in.)	0.25 mm (0.010 in.)	0.25 mm (0.010 in.)	0.30 mm (0.012 in.)
Exhaust	0.38 mm (0.015 in.)	0.25 mm (0.010 in.)	0.25 mm (0.010 in.)	0.38 mm (0.015 in.)
Valve Face Angle	46°	45°	45°	46°
Valve Seat Angle	46°	45°	45°	46°
Injection Timing, Static	....	26° BTDC	23° BTDC	....
Injector Opening Pressure	....	17235 kPa (2500 psi)	17235 kPa (2500 psi)	....
Ignition Timing, Static	12° BTDC	....	....	11° BTDC
Breaker-Point Gap	0.56 mm (0.022 in.)	....	....	0.56 mm (0.022 in.)
Dwell Angle	31°-34°	....	....	31°-34°
Spark Plug Gap	0.63 mm (0.025 in.)	....	....	0.63 mm (0.025 in.)
Governed Speeds — Engine Rpm				
Low Idle	725-775	725-775	725-775	725-775
High Idle (no-load)	2225-2275	2185	2185	2225-2275
Rated (full load)	2000	2000	2000	2000
Power Rating at Pto				
Shaft	37.3 kW (50 hp)	37.3 kW (50 hp)	38.8 kW (52 hp)	44.7 kW (60 hp)
SIZES-CLEARANCES				
Crankshaft Main Journal				
Diameter	76.162-76.175 mm (2.9985-2.9990 in.)	69.81-69.82 mm (2.7485-2.7490 in.)	76.162-76.175 mm (2.9985-2.9990 in.)	76.162-76.175 mm (2.9985-2.9990 in.)
Bearing Clearance	0.064-0.114 mm (0.0025-0.0045 in.)	0.076-0.127 mm (0.003-0.005 in.)	0.05-0.10 mm (0.002-0.004 in.)	0.064-0.114 mm (0.0025-0.0045 in.)

CONDENSED SERVICE DATA (CONT.)

	MF255 Gasoline	MF255 Diesel	MF255 Diesel	MF265 Gasoline
SIZES-CLEARANCES (CONT.)				
Crankshaft Crankpin				
Diameter	63.475-63.487 mm (2.4990-2.4995 in.)	57.112-57.125 mm (2.2485-2.2490 in.)	63.475-63.487 mm (2.4990-2.4995 in.)	63.475-63.487 mm (2.4990-2.4995 in.)
Bearing Clearance	0.038-0.076 mm (0.0015-0.0030 in.)	0.064-0.102 mm (0.0025-0.0040 in.)	0.038-0.076 mm (0.0015-0.0030 in.)	0.038-0.076 mm (0.0015-0.0030 in.)
Crankshaft End Play	0.05-0.38 mm (0.002-0.015 in.)	0.05-0.35 mm (0.002-0.014 in.)	0.10-0.38 mm (0.004-0.015 in.)	0.05-0.38 mm (0.002-0.015 in.)
Camshaft Journal Diameter				
Front	50.71-50.74 mm (1.9965-1.9975 in.)	47.47-47.50 mm (1.869-1.870 in.)	50.71-50.74 mm (1.9965-1.9975 in.)	50.71-50.74 mm (1.9965-1.9975 in.)
Center	50.46-50.48 mm (1.9865-1.9875 in.)	47.22-47.24 mm (1.859-1.860 in.)	50.46-50.48 mm (1.9865-1.9875 in.)	50.46-50.48 mm (1.9865-1.9875 in.)
Rear	49.95-49.97 mm (1.9665-1.9675 in.)	46.71-46.74 mm (1.839-1.840 in.)	49.95-49.97 mm (1.9665-1.9675 in.)	49.95-49.97 mm (1.9665-1.9675 in.)
Camshaft Bearing Clearance				
Front	0.064-0.132 mm (0.0025-0.0052 in.)	0.102-0.203 mm (0.004-0.008 in.)	0.064-0.132 mm (0.0025-0.0052 in.)	0.064-0.132 mm (0.0025-0.0052 in.)
Center & Rear	0.064-0.140 mm (0.0025-0.0055 in.)	0.102-0.203 mm (0.004-0.008 in.)	0.064-0.140 mm (0.0025-0.0055 in.)	0.064-0.140 mm (0.0025-0.0055 in.)
Camshaft End Play	0.10-0.40 mm (0.004-0.016 in.)	....	0.10-0.40 mm (0.004-0.016 in.)	0.10-0.40 mm (0.004-0.016 in.)
CAPACITIES				
Cooling System	10.4 L (11 U.S. qts.)	10.4 L (11 U.S. qts.)	10.4 L (11 U.S. qts.)	10.4 L (11 U.S. qts.)
Crankcase Oil	4.7 L* (5 U.S. qts.)	6.6 L* (7 U.S. qts.)	6.6 L* (7 U.S. qts.)	6.6 L* (7 U.S. qts.)
*Add 0.95 L (1 U.S. qt.) with filter change.				
Transmission, Differential and Hydraulic System				
With Dry Brakes	37.8 L (10 U.S. gals.)	37.8 L (10 U.S. gals.)	37.8 L (10 U.S. gals.)	37.8 L (10 U.S. gals.)
With Wet Brakes	41.6 L (11 U.S. gals.)	41.6 L (11 U.S. gals.)	41.6 L (11 U.S. gals.)	41.6 L (11 U.S. gals.)
Rear Axle Planetary (Each)	1.4 L (1.5 U.S. qts.)	1.4 L (1.5 U.S. qts.)	1.4 L (1.5 U.S. qts.)	1.4 L (1.5 U.S. qts.)
Power Steering	0.95 L (1 U.S. qt.)	0.95 L (1. U.S. qt.)	0.95 L (1 U.S. qt.)	0.95 L (1 U.S. qt.)
TIGHTENING TORQUES†				
Cylinder Head	122-129 N·m (90-95 ft.-lbs.)	95-102 N·m (70-75 ft.-lbs.)	135 N·m (100 ft.-lbs.)	122-129 N·m (90-95 ft.-lbs.)
Main Bearing Caps	230-244 N·m (170-180 ft.-lbs.)	150-155 N·m (110-115 ft.-lbs.)	230-244 N·m (170-180 ft.-lbs.)	230-244 N·m (170-180 ft.lbs.)
Connecting Rod Caps	————————Refer to Text————————			
Flywheel	100-108 N·m (74-80 ft.-lbs.)	100-108 N·m (74-80 ft.-lbs.)	108 N·m (80 ft.-lbs.)	100-108 N·m (74-80 ft.-lbs.)
Crankshaft Pulley	380-406 N·m (280-300 ft.-lbs.)	136-149 N·m (100-110 ft.-lbs.)	392 N·m (290 ft.-lbs.)	380-406 N·m (280-300 ft.-lbs.)
Rocker Shaft Supports	38-43 N·m (28-32 ft.-lbs.)	28-33 N·m (21-24 ft.-lbs.)	33 N·m (24 ft.-lbs.)	38-43 N·m (28-32 ft.-lbs.)
Intake Manifold	28-33 N·m (21-24 ft.-lbs.)	8-12 N·m (6-9 ft.-lbs.)	33 N·m (24 ft.-lbs.)	28-33 N·m (21-24 ft.-lbs.)
Exhaust Manifold	33-38 N·m (24-28 ft.-lbs.)	28-33 N·m (21-24 ft.-lbs.)	33 N·m (24 ft.-lbs.)	33-38 N·m (24-28 ft.-lbs.)
Camshaft Gear	61-68 N·m (45-50 ft.-lbs.)	26-28 N·m (19-21 ft.-lbs.)	65 N·m (48 ft.-lbs.)	61-68 N·m (45-50 ft.-lbs.)
Idler Gear Hub	27-33 N·m (20-24 ft.-lbs.)	26-28 N·m (19-21 ft.-lbs.)	41 N·m (30 ft.-lbs.)	26-28 N·m (20-24 ft.-lbs.)

†Torque figures apply with threads clean and lightly oiled.

CONDENSED SERVICE DATA

	MF265 Diesel	MF270 Diesel	MF275 Diesel	MF290 Diesel
GENERAL				
Engine Make	Perkins	Perkins	Perkins	Perkins
Engine Model.	A4.236	A4.236	A4.248	A4.248
Number of Cylinders	4	4	4	4
Bore	98.4 mm	98.4 mm	101 mm	101 mm
	(3.875 in.)	(3.875 in.)	(3.975 in.)	(3.975 in)
Stroke	127 mm	127 mm	127 mm	127 mm
	(5.0 in.)	(5.0 in.)	(5.0 in.)	(5.0 in)
Displacement	3.87 L	3.87 L	4.06 L	4.06 L
	(236 cu. in.)	(236 cu. in.)	(248 cu. in.)	(248 cu. in.)
Electrical System.	———————— 12 Volt, Negative Ground ————————			
Forward Speeds	———————————— 8 or 12 ————————————			
Reverse Speeds	———————————— 2 or 4 ————————————			
TUNE-UP				
Firing Order	1-3-4-2	1-3-4-2	1-3-4-2	1-3-4-2
Valve Clearance, Cold				
Intake and Exhaust	0.30 mm	0.30 mm	0.30 mm	0.30 mm
	(0.012 in.)	(0.012 in.)	(0.012 in.)	(0.012 in.)
Valve Clearance, Hot				
Intake and Exhaust	0.25 mm	0.25 mm	0.25 mm	0.25 mm
	(0.010 in.)	(0.010 in.)	(0.010 in.)	(0.010 in.)
Valve Face Angle.	45°	45°	45°	45°
Valve Seat Angle	45°	45°	45°	45°
Injection Timing, Static	23° BTDC	23° BTDC	24° BTDC	24° BTDC
Injector Opening Pressure	17235 kPa	17235 kPa	17235 kPa	17235 kPa
	(2500 psi)	(2500 psi)	(2500 psi)	(2500 psi)
Governed Speeds — Engine Rpm.				
Low Idle	725-775	725-775	725-775	725-775
High Idle (no-load).	2185	2185	2185	2380
Rated (full load)	2000	2000	2000	2200
Power Rating at Pto				
Shaft	44.7 kW	41.0 kW	50.0 kW	49.2 kW
	(60 hp)	(55 hp)	(67 hp)	(66 hp)
SIZES-CLEARANCES				
Crankshaft Main Journal				
Diameter	76.162-76.175 mm	76.162-76.175 mm	76.162-76.175 mm	76.162-76.175 mm
	(2.9985-2.9990 in.)	(2.9985-2.9990 in.)	(2.9985-2.9990 in.)	(2.9985-2.9990 in.)
Bearing Clearance	0.05-0.10 mm	0.05-0.10 mm	0.05-0.10 mm	0.05-0.10 mm
	(0.002-0.004 in.)	(0.002-0.004 in.)	(0.002-0.004 in.)	(0.002-0.004 in.)
Crankshaft Crankpin				
Diameter	63.475-63.487 mm	63.475-63.487 mm	63.475-63.487 mm	63.475-63.487 mm
	(2.4990-2.4995 in.)	(2.4990-2.4995 in.)	(2.4990-2.4995 in.)	(2.4990-2.4995 in.)
Bearing Clearance	0.038-0.076 mm	0.038-0.076 mm	0.038-0.076 mm	0.038-0.076 mm
	(0.0015-0.0030 in.)	(0.0015-0.0030 in.)	(0.0015-0.0030 in.)	(0.0015-0.0030 in.)
Crankshaft End Play	0.10-0.38 mm	0.10-0.38 mm	0.10-0.38 mm	0.10-0.38 mm
	(0.004-0.015 in.)	(0.004-0.015 in.)	(0.004-0.015 in.)	(0.004-0.015 in.)
Camshaft Journal Diameter				
Front	50.71-50.74 mm	50.71-50.74 mm	50.71-50.74 mm	50.71-50.74 mm
	(1.9965-1.9975 in.)	(1.9965-1.9975 in.)	(1.9965-1.9975 in.)	(1.9965-1.9975 in.)
Center	50.46-50.48 mm	50.46-50.48 mm	50.46-50.48 mm	50.46-50.48 mm
	(1.9865-1.9875 in.)	(1.9865-1.9875 in.)	(1.9865-1.9875 in.)	(1.9865-1.9875 in.)
Rear	49.95-49.97 mm	49.95-49.97 mm	49.95-49.97 mm	49.95-49.97 mm
	(1.9665-1.9675 in.)	(1.9665-1.9675 in.)	(1.9665-1.9675 in.)	(1.9665-1.9675 in.)
Camshaft Bearing Clearance				
Front	0.064-0.132 mm	0.064-0.114 mm	0.64-0.132 mm	0.064-0.114 mm
	(0.0025-0.0052 in.)	(0.0025-0.0045 in.)	(0.0025-0.0052 in.)	(0.0025-00.0045 in.)
Center & Rear	0.064-0.140 mm	0.064-0.140 mm	0.064-0.140 mm	0.064-0.140 mm
	(0.0025-0.0055 in.)	(0.0025-0.0055 in.)	(0.0025-0.0055 in.)	(0.0025-0.0055 in.)
Camshaft End Play	0.10-0.40 mm	0.10-0.40 mm	0.10-0.40 mm	0.10-0.40 mm
	(0.004-0.016 in.)	(0.004-0.016 in.)	(0.004-0.016 in.)	(0.004-0.016 in.)

CONDENSED SERVICE DATA (CONT.)

	MF265 Diesel	MF270 Diesel	MF275 Diesel	MF290 Diesel
CAPACITIES				
Cooling System	10.4 L	14.2 L	11.3 L	14.2 L
	(11 U.S. qts.)	(15 U.S. qts.)	(12 U.S. qts.)	(15 U.S. qts.)
Crankcase Oil..................	6.6 L*	6.6 L*	6.6 L*	6.6 L*
	(7 U.S. qts.)	(7 U.S. qts.)	(7 U.S. qts.)	(7 U.S. qts.)
*Add 0.95 L (1 U.S. qt.) with filter change.				
Transmission, Differential and Hydraulic System				
With Dry Brakes	37.8 L	....	37.8 L	....
	(10 U.S. gals.)		(10 U.S. gals.)	
With Wet Brakes	41.6 L	41.6 L	41.6 L	41.6 L
	(11 U.S. gals.)	(11 U.S. gals.)	(11 U.S. gals.)	(11 U.S. gals.)
Rear Axle Planetary (Each)	1.4 L	1.4 L	1.4 L	1.4 L
	(1.5 U.S. qts.)	(1.5 U.S. qts.)	(1.5 U.S. qts.)	(1.5 U.S. qts.)
Power Steering	0.95 L	1.1 L	0.95 L	1.1 L
	(1 U.S. qt.)	(1.2 U.S. qt.)	(1 U.S. qt.)	(1.2 U.S. qt.)
TIGHTENING TORQUES†				
Cylinder Head	135 N·m	135 N·m	135 N·m	135 N·m
	(100 ft.-lbs.)	(100 ft.-lbs.)	(100 ft.-lbs.)	(100 ft.-lbs.)
Main Bearing Caps	244 N·m	244 N·m	244 N·m	244 N·m
	(180 ft.-lbs.)	(180 ft.-lbs.)	(180 ft.-lbs.)	(180 ft.-lbs.)
Connecting Rod Caps	————————————Refer to Text————————————			
Flywheel	108 N·m	108 N·m	108 N·m	108 N·m
	(80 ft.-lbs.)	(80 ft.-lbs.)	(80 ft.-lbs.)	(80 ft.-lbs.)
Crankshaft Pulley	392 N·m	392 N·m	392 N·m	392 N·m
	(290 ft.-lbs.)	(290 ft.-lbs.)	(290 ft.-lbs.)	(290 ft.-lbs.)
Rocker Shaft Supports	33 N·m	33 N·m	33 N·m	33 N·m
	(24 ft.-lbs.)	(24 ft.-lbs.)	(24 ft.-lbs.)	(24 ft.-lbs.)
Intake Manifold	33 N·m	33 N·m	33 N·m	33 N·m
	(24 ft.-lbs.)	(24 ft.-lbs.)	(24 ft.-lbs.)	(24 ft.-lbs.)
Exhaust Manifold	33 N·m	33 N·m	33 N·m	33 N·m
	(24 ft.-lbs.)	(24 ft.-lbs.)	(24 ft.-lbs.)	(24 ft.-lbs.)
Camshaft Gear................	61-68 N·m	65 N·m	61-68 N·m	65 N·m
	(45-50 ft.-lbs.)	(48 ft.-lbs.)	(45-50 ft.-lbs.)	(48 ft.-lbs.)
Idler Gear Hub	27-32 N·m	40 N·m	27-32 N·m	40 N·m
	(20-24 ft.-lbs.)	(30 ft.-lbs.)	(20-24 ft.-lbs.)	(30 ft.-lbs.)

†Torque figures apply with threads clean and lightly oiled.

FRONT SYSTEM

AXLE ASSEMBLY

All Models

1. Refer to Fig. 1 for an exploded view of adjustable axle typical of unit used on all tractors. Recommended toe-in is 3.2 mm (1/8 inch) and is adjusted by turning right-hand tie rod end (E) into or out of tube (T).

To remove the axle assembly, first remove hood, grille, side panels, battery, battery support, shield and grille frame. Remove snap ring (5) and pull the center steering shaft (11) down out of arm (2) and support (1). Be careful not to lose shims (6) or damage seal (9). Support axle and engine separately. Remove retaining clips, then remove retaining pin (14). Use a puller to withdraw pivot pin (16) out toward front. Fore and aft play of axle is adjusted by thickness of shims (15) and washer (19).

To remove axle support (1), first remove radiator and power steering cylinder. Remove mounting bolts, then lower support housing from tractor.

Inspect all parts for wear or damage and renew if necessary. Ream new bushings after installation to provide desired operating clearance.

To reinstall axle support and axle assembly, reverse the removal procedure. Tighten cap screws retaining axle support casting to 271 N·m (200 ft.-lbs.) torque. Add shims (15) as necessary to limit fore and aft play of axle to 0.08-0.25 mm (0.003-0.010 inch). Be sure shims are to the front and thrust washer (19) is to the rear.

SPINDLES AND WHEEL HUBS

All Models

2. It is recommended that front wheel bearings be cleaned, inspected and repacked with grease after every 500 hours of operation, or annually, whichever comes first. A good quality multipurpose lithium base grease is recommended for repacking bearings.

To remove spindle (22—Fig. 1), loosen

spindle arm (13) clamp bolt and pull the arm from spindle shaft. Remove key from spindle shaft, then lower spindle from axle extension (25).

Inspect thrust bearing (27) and bushings (24 and 26) and renew if necessary. Spindle bushings must be reamed after installation to provide desired shaft to bushing clearance. Lubricate with multipurpose lithium base grease.

Reinstall spindle and arm with a new dust seal (23). Position arm on spindle so end play is less than 0.10 mm (0.004 inch), then tighten clamp bolt to 135 N·m (100 ft.- lbs.) torque.

To adjust wheel hub bearings, tighten wheel retaining nut to 80 N·m (60 ft.-lbs.) torque while rotating wheel hub. Back off castellated nut to align nearest slot with pin hole, then back off further to the next slot and install cotter pin.

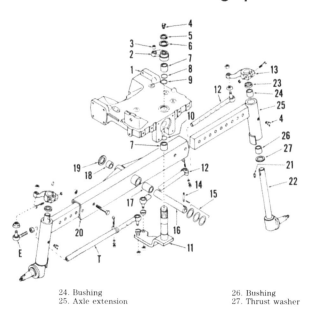

Fig. 1—Exploded view of typical adjustable front axle assembly used on all models. A two-piece steering arm and shaft (11) is used on some models.

1. Front support
2. Arm
3. Bushing
4. Grease fittings
5. Snap ring
6. Shim
7. Bushings
8. "O" ring
9. Seal
10. "O" ring
11. Steering arm & shaft
12. Tie rod
13. Spindle arm
14. Retaining pin
15. Shims
16. Axle pivot pin
17. Bushing
18. Bushing
19. Washer
20. Axle
21. Woodruff key
22. Spindle
23. Seal
24. Bushing
25. Axle extension
26. Bushing
27. Thrust washer

POWER STEERING SYSTEM

LUBRICATION AND BLEEDING

All Models

3. The hydrostatic power steering fluid reservoir is attached to the rear of the power steering hydraulic pump (Fig. 2). Massey-Ferguson Permatran 111 Oil is recommended for use in steering system.

On early models, an oil strainer screen is located on pump inlet pipe inside the reservoir. On late models. a renewable oil filter element is located within the reservoir. On all models, it is recommended that steering fluid be renewed and suction screen cleaned, or filter element replaced, after every 500 hours of operation, or annually, whichever comes first.

The steering system is self bleeding, but steering should be cycled and reservoir refilled as necessary until level stops dropping. To fill, stop engine and remove filler plug (F—Fig. 2). If steering cylinder has been disassembled or drained, fill reservoir then start and idle engine, adding fluid as level lowers until system is stabilized. Install filler plug loosely, cycle the system then recheck, adding fluid as necessary to maintain full reservoir. Tighten plug securely when fluid level ceases to drop.

OPERATING PRESSURE

All Models

4. To check power steering relief valve pressure, first start engine and actuate steering until oil temperature is approximately 50° C (120° F). Disconnect pressure line (P—Fig. 2) and connect a 0-20000 kPa (0-3000 psi) pressure gage in pump outlet port. Start engine and operate at 2000 rpm. Relief pressure should not exceed 11375 kPa (1650 psi).

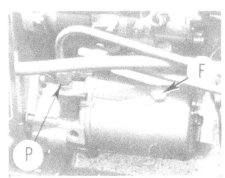

Fig. 2—The power steering reservoir is filled through hole for plug (F). Pump pressure line is shown at (P).

Fig. 3—Exploded view of Parker-Hannifin power steering pump. Refer to text for assembly notes.

1. Drive gear
2. Seal
3. Front cover
4. Seals
8. Pumping gears
9. Pump body
10. Rear cover
11. Relief valve
13. Filter
14. Reservoir
15. "V" seal
16. Gasket
17. Shield
18. Wear plate
19. Thrust plate
20. Screws

Relief valve pressure setting can be adjusted by turning relief valve adjusting screw (Fig. 3, 4 or 5) clockwise to increase pressure or counterclockwise to reduce pressure. On early style pump (Fig. 3) reservoir (14) must be removed for access to relief valve. On late style pumps (Fig. 4 or 5), relief valve is located on outside of pump body. Recommended pressure setting is 10345-11030 kPa (1500-1600 psi).

If specified relief pressure cannot be obtained, service relief valve and pump as outlined in appropriate paragraph 5, 6 or 7.

POWER STEERING PUMP

Three different types of power steering pumps have been used; Parker-Hannifin, Sundstrand and Aero-Quip. See Figs. 3, 4 and 5 for exploded view of each pump. The gear type steering pump is attached to rear of engine tim-

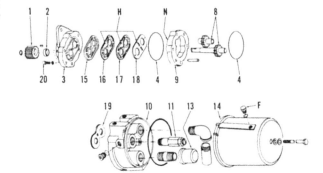

ing gear case, on either the left-hand or right-hand side of engine (depending on tractor model), and is driven by the timing gears.

Parker-Hannifin Pump

5. R&R AND OVERHAUL. Be sure to thoroughly clean outside of pump and hydraulic lines before removing pump from engine. The pump mounting cap screws enter pump housing from the front and heads are located on timing gear front cover.

Before disassembling, scribe alignment marks across front cover (3—Fig. 3), body (9), rear cover (10) and reservoir (14). Separate reservoir from rear cover. Remove drive gear (1) and key from pump shaft. Remove tie bolts, then tap end of pump drive shaft to separate rear cover from front cover. Place alignment marks on teeth of pump gears so original contact pattern of gear teeth can be maintained if original gears are reinstalled. Remove gears, thrust plate (19) and wear plate (18) from pump body. Remove shield (17), gasket (16), "V" seal (15) and drive shaft seal (2) from front cover.

If relief valve (11) is to be disassembled, count the number of turns required to remove the adjusting plug so plug can be reinstalled to its original setting. Check relief pressure as outlined in paragraph 4 after pump is reinstalled and adjust if necessary.

Inspect all parts for evidence of wear or damage. If pump body (9) is excessively worn, pump assembly should be renewed. Renew all seals, gaskets, thrust plate, wear plate and "O" rings.

To reassemble, reverse the disassembly procedure while noting the following special instructions: Lip of oil seal (2) should face inward in front cover. The "V" groove of seal (15) should be downward (facing away from gasket) in front cover. The 1.5 mm (1/16 inch) hole (H) in gasket (16), shield (17) and wear plate (18) must be positioned upward (toward pressure side of pump) with bronze side of wear plate toward gear faces. Notch (N) on flat surface of pump body must be over holes (H). Bronze side of thrust plate (19) must be toward gears with large cut-out section toward inlet (bottom) side of pump. Make certain all assembly marks on pump gears and outside of pump body are aligned, then tighten pump assembly cap screws evenly to 27 N·m (20 ft.-lbs.) torque. Tighten reservoir mounting bolt to 11.3 N·m (100 in.-lbs.) torque.

If pump is mounted on left side of engine, counterbore on drive gear (1) should face away from pump. If pump is mounted on right side of engine, install gear with counterbore toward

pump. Reinstall pump and connect hydraulic lines. Fill reservoir with oil, start engine and cycle steering from side to side several times to purge air from system. Check relief valve pressure setting as outlined in paragraph 4.

Sundstrand Pump

6. R&R AND OVERHAUL. Thoroughly clean exterior of pump and surrounding area before removing. Disconnect hydraulic lines, remove attaching cap screws and withdraw pump assembly.

Prior to disassembly, scribe alignment marks across pump body, end plates and reservoir. Remove reservoir (15—Fig. 4) and filter element (14). Remove drive gear and key from pump drive shaft. Remove tie bolts, then carefully separate end plates (3 and 12) from pump body (9). Remove pump gears (7 and 8) and bearings (6). Remove seal rings (4 and 5) and shaft oil seal (2). Remove relief valve assembly (11).

Inspect all parts for evidence of wear or damage and renew as necessary. Note that the pump gears are loaded toward inlet side of pump body (9) due to hydraulic pressure within pump, thus gears will normally cut a track on inlet side of pump body if pump gear shafts and/or bearings are worn. If depth of wear track exceeds 0.10 mm (0.004 inch), pump assembly should be renewed. Overall thickness of gears and bearings should be 0.10-0.20 mm (0.004-0.008 inch) less than thickness of pump body.

Use new seals and "O" rings when reassembling. Be sure that relieved side of bearings (6) is located on outlet side of pump. Tighten body retaining bolts evenly to 41 N·m (30 ft.-lbs.) torque. Be sure scribe marks made prior to disassembly are aligned.

Reinstall pump and tighten mounting cap screws to 27 N·m (20 ft.-lbs.) torque. Connect hydraulic lines, refill reservoir and check relief valve pressure as outlined in paragraph 4.

Aero-Quip Pump

7. R&R AND OVERHAUL. To remove pump, first thoroughly clean exterior of pump and surrounding area. Disconnect hydraulic lines, remove attaching cap screws and withdraw pump assembly.

Before disassembling pump, scribe alignment marks across pump body, end plates and reservoir to ensure correct reassembly. Remove reservoir (17—Fig. 5) and filter element (16). Remove drive gear (2) and Woodruff key from pump drive shaft. Remove tie bolts, then carefully separate end plates (5 and 14) from pump body (13). Remove pump gears (8) and bearings (9). Remove relief valve assembly (19).

Inspect all parts for evidence of wear or damage and renew if necessary. A seal repair kit is available for renewal of all seals and "O" rings.

To reassemble, reverse the disassembly procedure. Tighten body retaining bolts evenly while making certain that pump drive shaft turns freely.

Reinstall pump and tighten mounting

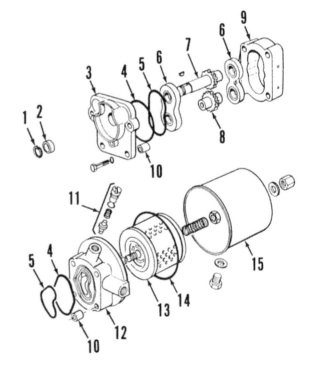

Fig. 4—Exploded view of Sundstrand power steering pump used on some models.

1. Snap ring
2. Oil seal
3. Front cover
4. "O" ring
5. Seal ring
6. Bearings
7. Drive gear
8. Driven gear
9. Pump body
10. Spacers
11. Relief valve assy.
12. Rear cover
13. Filter element
14. "O" ring
15. Reservoir tank

cap screws to 27 N·m (20 ft.-lbs). Connect hydraulic lines, refill reservoir and check relief valve pressure as outlined in paragraph 4.

HYDROSTATIC HAND PUMP

Early Models

8. R&R AND OVERHAUL. Remove cover from left-hand side of instrument panel. Disconnect the hydraulic lines from hydrostatic hand pump and plug all openings to prevent entry of dirt. Remove the four cap screws attaching hand pump to mounting bracket and steering column, then withdraw unit from splined end of steering shaft (Fig. 7).

To disassemble, first thoroughly clean exterior of control unit. Remove cap screws attaching metering pump to bottom of housing, then remove end cap (22—Fig. 6), rotor assembly (21), spacer plate (20) and drive link (18) from valve body. Remove screws retaining top cover (7) and lift off cover. Use a brass rod with one end bent 90° to push check valve plug from valve body as shown in Fig. 8. Remove check valve seat using an Allen wrench, then remove check valve ball and spring (Fig. 9). Carefully push control valve spool and sleeve assembly out bottom of housing as shown in Fig. 10.

NOTE: Be careful valve unit does not bind. Parts are fit to extremely close tolerance and a twisting motion may be required for withdrawal.

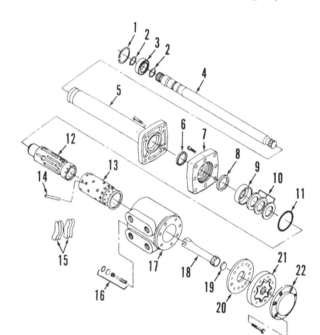

Fig.6—Exploded view of hydrostatic steering hand pump and control valve assembly used on some early model tractors.

1. Snap ring
2. Snap rings
3. Bearing
4. Steering wheel shaft
5. Steering column
6. Oil seal
7. End cover
8. Locator bushing
9. Bushing
10. Thrust bearing assy.
11. "O" ring
12. Valve spool
13. Valve sleeve
14. Drive pin
15. Centering springs
16. Check valve assy.
17. Valve body
18. Drive link
19. Disc
20. Spacer plate
21. Rotor assy.
22. End cap

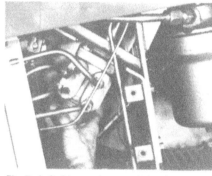

Fig. 7—Left side panel removed showing removal of hydrostatic hand pump.

Fig. 8—Use a brass welding rod with a 90° bend in one end to form a leg about 8 mm (5/16 inch) long to push check valve plug from valve body as shown.

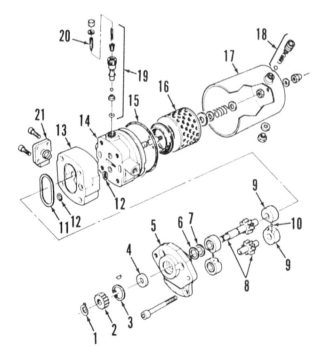

Fig. 5—Exploded view of Aero-Quip power steering pump used on some models.

1. Snap ring
2. Drive gear
3. Snap ring
4. Oil seal
5. Front cover
6. Thrust washer
7. "O" ring
8. Pump gears
9. Bearings
10. Centering pin
11. Seal ring
12. "O" ring
13. Pump body
14. Rear cover
15. "O" ring
16. Filter element
17. Reservoir tank
18. Breather assy.
19. Relief valve assy.
20. Adjusting screw
21. Port plate

Fig. 9—View of check valve components removed from valve body.

Remove drive pin, then separate spool and sleeve as shown in Fig. 11. Remove centering springs from spool.

Thoroughly clean all parts in suitable solvent and blow dry with compressed air. Inspect lapped surfaces for scoring or other damage and renew parts if necessary. It is permissible to hand lap light scratches or burrs from mating surfaces of end cap (22—Fig. 6), rotor assembly (21), spacer plate (20) and valve body (17). Use 600 grit abrasive paper or lapping compound on a surface plate or piece of plate glass. Rinse in clean solvent and blow dry after polishing is complete. Rotor unit (21) and the valve body, spool (12) and sleeve (13) are available only as complete assemblies. Renew oil seal and all "O" rings.

To reassemble, reverse the disassembly procedure while noting the following special instructions: Tighten check valve seat to a torque of 17.6 N·m (155 in.-lbs.). Install centering springs in spool in sets of three with arches toward the middle as shown in Fig. 12. Assemble valve spool into sleeve so centering springs are at slotted end of sleeve. Insert drive pin. Lubricate parts with clean hydraulic oil, then install spool and sleeve into bottom end of valve body. Be sure that pin slot in end of drive link is aligned with valley of inner rotor as shown in Fig. 14. This will correctly "time" pump metering element to valve spool assembly. Be sure that slot in drive link engages drive pin in valve spool. Disc (19—Fig. 6) will be flush with surface of rotor set (21) when drive link is correctly installed. Install end cap (22) and tighten retaining cap screws evenly to 28 N·m (21 ft.-lbs.) torque. Install locator bushing (8) with chamfer on outside diameter facing up. Install top cover (7) and tighten retaining cap screws evenly to 26 N·m (19 ft.-lbs.) torque.

Reinstall assembled unit and bleed air from system as outlined in paragraph 3.

Late Models

9. R&R AND OVERHAUL. Remove lower cover from instrument panel. Note position of hydraulic pipes connected to hand pump to ensure correct reassembly. Disconnect hydraulic pipes and immediately plug all openings to prevent entry of dirt. Remove the four bolts attaching hand pump unit to support bracket and steering column. Remove cap screw retaining steering column to instrument panel, raise the steering column for clearance and remove hand pump from tractor.

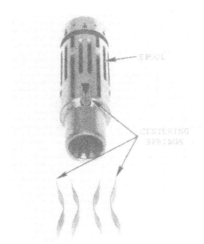

Fig. 12—View showing correct installation of centering springs (arch to arch in sets of three, notched edge down).

Prior to disassembly, thoroughly clean the exterior of hand pump unit. Remove screws retaining end plate (21—Fig. 15), then lift off end plate, stator and rotor set (20), spacer (19), valve plate (17) and drive link (16). Remove threaded insert and ball (7). Slide spool (15) and sleeve (11) as an assembly from valve body. Remove locating bushing (12), ring (10), if so equipped, thrust bearing assembly (9), "O" ring (8) and Teflon ring (6). Remove oil seal (2) from top of valve body. Push the drive pin (14) out of sleeve and spool, then slide the spool from the sleeve. Push the centering springs (13) out of the slot in valve spool.

The steering shock valve (3) is set at the factory to open at a pressure of 20000 kPa (2900 psi). Removal of shock valve is not recommended as a suitable hydraulic test stand is required to properly set opening pressure of valve.

Thoroughly clean all parts in a suitable solvent being careful not to scratch or nick polished surfaces of pump and control valve components. Inspect all parts for wear or damage and renew if necessary. If either the spool, sleeve or valve body is scored or damaged, the steering valve must be renewed as a complete unit.

If lapped surfaces of valve body (1), valve plate (17), stator/rotor set (20) or end plate (21) are scored or nicked, they should be renewed. However, it is permissible to hand lap very light scratches or burrs from these parts if desired. Use 600 grit abrasive paper or lapping compound on a surface plate or piece of plate glass. Rinse parts in clean solvent and blow dry after polishing is completed. A valve repair kit is available which contains gaskets, seals and "O" rings necessary for reassembly of unit.

Lubricate parts with clean hydraulic oil during reassembly. To reassemble, reverse the disassembly procedure while noting the following special instructions: Assemble centering springs into slot in valve sleeve and spool in sets of three with arches toward the middle as shown in Fig. 12. Insert drive pin

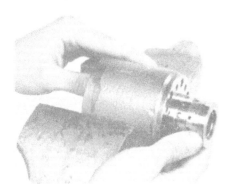

Fig. 10—Push spool and sleeve assembly out bottom of housing as shown.

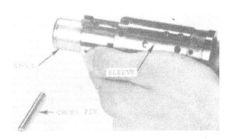

Fig. 11—Spool can be pushed from sleeve after removing cross pin.

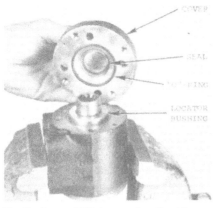

Fig. 13—Be sure "O" rings, seal and bushing are properly installed when reassembling.

Fig. 14—When properly assembled, slot in end of drive link will align with a valley of inner rotor as shown.

(14—Fig. 15), then carefully slide valve spool assembly into bottom of valve body. Be sure that pin slot in end of drive link is aligned with valley of inner rotor as shown in Fig. 14. This will

correctly "time" pump metering element to valve spool assembly. Make sure that slot in drive link engages drive pin in valve spool. Be sure that dowel pin (22) engages threaded check ball insert (7). Tighten end cover retaining cap screws evenly to a torque of 27 N·m (20 ft.-lbs.).

To reinstall steering hand pump, reverse the removal procedure. Bleed air from system as outlined in paragraph 3.

STEERING CYLINDER

All Models

10. R&R AND OVERHAUL. To remove steering cylinder, first remove grille, side panels, battery, battery platform and shield. Disconnect hoses from cylinder, then plug all openings to prevent entry of dirt. Pull pivot pin from end of cylinder rod. Disconnect the grease supply pipe (if so equipped). Remove clamp bolt (if so equipped) and snap ring retaining steering arm to top of steering shaft. Remove cylinder and steering arm as a unit.

Two types of steering cylinders have been used (refer to Fig. 16 and Fig. 17). Disassembly of cylinder shown in Fig. 16 is evident after removing the four tie bolts (4).

To disassemble cylinder shown in Fig. 17, turn bearing cap (3) until straight end of retaining wire (14) is positioned in hole (H) of cylinder barrel. Use a small screwdriver or similar tool to start retainer wire out of the hole, then rotate bearing cap to force wire out of hole. After retainer wire is removed, pull rod (8), piston (20) and bearing cap (3) out of cylinder (7).

On either type cylinder, check parts for wear, scoring or other damage and renew if necessary. Renew all seals, "O" rings and back-up rings. Warming Teflon

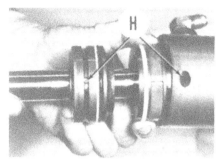

Fig. 18—Holes (H) are used to remove and install retaining wire (14—Fig. 17). Refer to text.

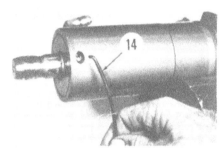

Fig. 19—The hooked end of retaining wire is inserted into hole in bearing cap and cap is turned to pull wire into the groove.

Fig. 20—Turn the bearing cap the direction indicated by arrow to pull retaining wire into groove. Use a screwdriver to start end of wire out of hole and turn end cap opposite direction shown to remove wire.

Fig. 15—Exploded view of hydrostatic steering hand pump and control valve assembly used on late production tractors.

1. Valve housing	13. Centering springs
2. Seal	14. Drive pin
3. Shock relief valve	15. Valve sleeve
4. Ball & plunger	16. Drive link
5. "O" ring	17. Valve plate
6. Teflon ring	18. "O" rings
7. Ball & plug	19. Spacer
8. "O" ring	20. Stator/rotor set
9. Thrust bearing assy.	21. End plate
10. Ring	22. Dowel pin
11. Valve spool	23. Drilled cap screw
12. Locator bushing	

Fig. 16—Exploded view of one of the two types of steering cylinders. Refer to Fig. 17 for other type.

1. Seal
2. Seal
3. Bearing cap
4. Assembly screws
5. "O" ring
6. Back-up ring
7. Cylinder
8. Rod & piston
9. Piston ring

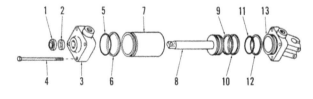

10. Piston ring
11. "O" ring

12. Back-up ring
13. End cap

Fig. 17—Exploded view of one type of steering cylinder. Angle drilled hole (H) is also shown in Fig. 18.

1. Seal
2. Seal
3. Bearing cap
7. Cylinder
8. Rod
14. Retaining wire
15. Back-up ring
16. "O" ring

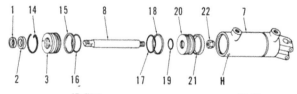

17. Ring
18. Ring
19. "O" ring

20. Piston
21. Piston ring
22. Nut

piston ring in hot oil will make installation over piston easier. Allow seal ring to cool and return to its original shape before assembling into cylinder. Lubricate "O" rings and seals with clean hydraulic oil during reassembly.

When reinstalling cylinder, be sure to install shims (if used) with steering arm. Bleed system as outlined in paragraph 3.

PERKINS GASOLINE ENGINE AND COMPONENTS

Model MF255 may be equipped with a Perkins AG4.212 four cylinder gasoline engine. Model MF265 may be equipped with a Perkins AG4.236 four cylinder gasoline engine. Service procedures are similar for both engines.

R&R ENGINE ASSEMBLY

All Gasoline Models

11. To remove the engine and clutch as a unit, first drain cooling system, and if engine is to be disassembled, drain oil pan. Remove front grille panels, disconnect battery cables and remove hood. Disconnect headlight wires, radiator hoses, air cleaner hose and hydraulic oil cooler lines. Detach battery cable from starter motor, remove left rear side panel and disconnect power steering oil lines. Support tractor with jack at front of transmission and install wedges between front axle and axle support to prevent tipping. Support front axle so that it can be moved forward, then unbolt and separate front assembly from engine.

Shut fuel off at tank, disconnect fuel level gage sending unit wire and fuel line then remove tank, support plate and heat shield. Disconnect main wiring harness from alternator, starter, temperature gage sending unit and ignition coil. Disconnect oil pressure line, tachometer drive cable, power steering pump oil lines and choke cable. Remove muffler from models with vertical exhaust, or disconnect exhaust pipe on models with low exhaust. On all models, remove throttle rod and be sure that all wires, linkage and lines are positioned out of the way for engine removal. Attach lifting eye to left front of cylinder head and support engine with hoist. Remove screws attaching engine to transmission housing. Use guide studs when separating engine from transmission.

Reinstall in reverse of removal procedure. Coat input shaft splines lightly with lithium base grease before assembling.

CYLINDER HEAD

All Gasoline Models

12. REMOVE AND REINSTALL. To remove the cylinder head, drain cooling system and remove hood. Disconnect fuel gage sending unit wire and fuel line then remove fuel tank, support plate and heat shield. On models with vertical exhaust, remove muffler; on models with low exhaust, disconnect exhaust pipe. On all models, disconnect upper radiator hose, air cleaner hose, alternator bracket, spark plug wires, governor linkage and breather tube. Disconnect wires from fuel shut-off solenoid, ignition coil and temperature gage sending unit. Remove ignition coil and fuel line. Remove the carburetor, intake manifold and exhaust manifold. Remove valve cover, rocker arm assembly, push rods and spark plugs. Loosen cylinder head retaining cap screws and stud nuts in reverse of order shown in Fig. 21.

Check cylinder head surface for distortion, cracks or other damage. Maximum allowable warpage is 0.10 mm (0.004 inch). Cylinder head may be resurfaced providing specified valve head location relative to cylinder head surface is maintained (refer to paragraph 13). Original cylinder head thickness is 102.81-103.57 mm (4.048-4.077 inches).

Clean mating surfaces of block and head, then position gasket with "TOP FRONT FIT DRY" toward top front. Install gasket dry, do no use sealing compound. Install cylinder head and tighten fasteners in sequence shown in Fig. 21 evenly in 15-20 N·m (10-15 ft.-lbs.) increments until final torque of 122-129 N·m (90-95 ft.-lbs.) is obtained. Install push rods and tighten rocker arm shaft supports to 38-43 N·m (28-32 ft.-lbs.) torque. Remainder of assembly is reverse of removal procedure. Retorque cylinder head and adjust valve clearance after engine has reached normal operating temperature. Valve clearance (tappet gap) should be 0.30 mm (0.012 inch) for inlet valves; 0.38 mm (0.015 inch) for exhaust valves.

VALVES AND SEATS

All Gasoline Models

13. Intake valves seat directly in cylinder head and exhaust valves seat on renewable valve seat inserts. Face and seat angle for all valves should be 46°. The intake valve seat has a 31° flare machined between seat and cylinder head face. Minimum intake valve seat contact width is 1.83 mm (0.072 inch) and minimum recommended exhaust valve seat contact width is 2.82 mm (0.110 inch). Renew valve if width of valve head margin is less than 0.8 mm (1/32 inch) after grinding valve face.

Check valve head location relative to surface of cylinder head as shown in Fig. 22. Intake valve must be recessed BELOW cylinder head surface within limits of 0.91-1.14 mm (0.036-0.045 inch). Exhaust valve must stand out ABOVE cylinder head surface within limits of 1.57-2.06 mm (0.062-0.081 inch).

Fig. 22—Intake valve should be recessed 0.91-1.14 mm (0.036-0.045 inch) and exhaust should be raised 1.57-2.06 mm (0.062-0.081 inch) above gasket surface of head.

Fig. 21—On Perkins gasoline engines, tighten cylinder head fasteners evenly to a torque of 122-129 N·m (90-95 ft.-lbs.) in the sequence shown.

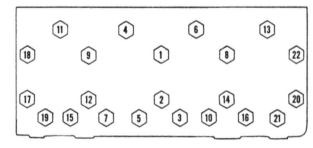

The exhaust valve seat insert can be renewed. Depth of service valve seat insert is 0.25 mm (0.010 inch) less than production insert which necessitates machining cylinder head gasket surface 0.25 mm (0.010 inch) before installing new seats. Depth of counterbore should be 6.35-6.48 mm (0.250-0.255 inch) before resurfacing head to reduce depth to 6.10-6.22 mm (0.240-0.245 inch). Diameter of cylinder head counterbore should be 0.05-0.08 mm (0.002-0.003 inch) smaller than insert diameter to provide an interference fit. Shrink insert in dry ice, then press fully into counterbore. Peen cylinder head around insert to prevent insert from loosening. Reface valve seat after installation. Be sure to check for correct valve stand out.

Standard valve stem diameter is 9.4615-9.4869 mm (0.3725-0.3735 inch) for intake valves and 9.4488-9.4678 mm (0.3720-0.3727) for exhaust valves. Valves with oversize valve stems are also available to provide correct valve stem to guide clearance. Special oversize valve stem seals must be used with 0.38 mm (0.015 inch) and 0.76 mm (0.030 inch) oversize valve stems.

Valve springs and keepers are alike for both intake and exhaust valves. Valve rotators are installed only on exhaust valves and stem seals are installed only on intake valves. Install all valve springs with closely wound coils toward cylinder head.

VALVE GUIDES

All Gasoline Models

14. Intake and exhaust valve guides are cast into the cylinder head, and valves with oversize stems are available to provide correct stem to guide clearance. Standard bore of all valve guides is 9.50595-9.51865 mm (0.37425-0.37475 inch). Stem to guide clearance should be 0.01905-0.05715 mm (0.00075-0.00225 inch) for intake; 0.03810-0.06985 mm (0.00150-0.00275 inch) for exhaust. Be sure to use oversize valve stem seal on valves that are 0.38 mm (0.015 inch) and 0.76 mm (0.030 inch) oversize. Standard size stem seal can be used on 0.08 mm (0.003 inch) oversize valve stem. Be sure to reseat valves after resizing valve guides.

VALVE SPRINGS

All Gasoline Models

15. Valve springs are alike for intake and exhaust valves. Springs should be installed with close wound coils toward cylinder head. Renew any spring which

is discolored, distorted or fails to meet the following specifications:

Outer Spring
Test Length . . . . .45.2 mm
(1-23/32 in.)
Test Load . . . . .169-186 N
(38-42 lbs.)
Inner Spring
Test Length . . . . .39.69 mm
(1-9/16 in.)
Test Load . . . . .66-71 N
(14.7-16.1 lbs.)

ROCKER ARMS

All Gasoline Models

16. The rocker arms and shaft assembly can be removed after removing hood, fuel tank and rocker arm cover. Rocker arms are right-hand and left-hand units which must be installed on shaft as shown in Fig. 23.

Desired diametral clearance between new rocker arm bushings and new shaft is 0.025-0.089 mm (0.0010-0.0035 inch). Bushings must be reamed to 19.063-19.101 mm (0.7505-0.7520 inch) after installation. Be sure to align hole in rocker arm and hole in bushing. Renew shaft and/or bushing if clearance is excessive. Tighten rocker arm shaft support retaining stud nuts to 38-43 N·m (28-32 ft.-lbs.) torque. Adjust valve clearance as outlined in paragraph 17 after rocker arm assembly is reinstalled.

VALVE CLEARANCE

All Gasoline Models

17. Recommended valve clearance (tappet gap) after engine reaches normal operating temperature is 0.30 mm (0.012 inch) for intake valves, 0.38 mm (0.015 inch) for exhaust valves. All valves can be adjusted from just two crankshaft positions using Figs. 25 and 26 and the following procedure.

Turn crankshaft until TDC timing mark on crankshaft pulley aligns with timing pointer as shown in Fig. 24 and check rocker arms for front and rear cylinders. If rear rocker arms are tight

and front rocker arms loose, No. 1 piston is on compression stroke; adjust the valves indicated on Fig. 25 first. Turn crankshaft one complete turn until timing mark again aligns and adjust remainder of valves indicated in Fig. 26. If No. 4 piston is on compression stroke (front rocker arms tight), adjust the four valves indicated in Fig. 26 first, then

Fig. 24—Timing marks are located on crankshaft pulley as shown.

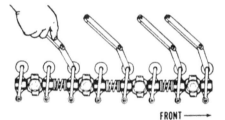

Fig. 25—With TDC timing marks aligned as shown in Fig.24 and No. 1 piston on compression stroke, adjust the indicated valves. Turn crankshaft one complete turn, refer to Fig. 26 for remainder of valves.

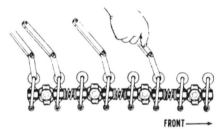

Fig. 26—With TDC timing marks aligned as shown in Fig.24 and No. 1 piston on exhaust stroke, adjust the indicated valves.

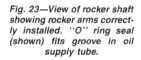

Fig. 23—View of rocker shaft showing rocker arms correctly installed. "O" ring seal (shown) fits groove in oil supply tube.

turn crankshaft one complete turn and adjust remainder of valves.

TIMING GEAR COVER

All Gasoline Models

18. To remove the timing gear cover, first drain the cooling system, remove front grille panels, disconnect battery cables and remove hood. Disconnect headlight wires, radiator hoses, air cleaner hose and hydraulic oil cooler lines. Detach battery cable from starter motor, remove left rear side panel and disconnect power steering lines. Support tractor with a jack at front of transmission and place wedges between front axle and axle support to prevent tipping. Support front axle so it can be moved forward, then unbolt and separate front assembly from engine.

Remove fan belt, fan blades and crankshaft pulley. Unbolt and remove timing gear cover being careful not to bend cover.

Crankshaft front oil seal can be renewed at this time. Install seal with lip to rear and with front edge of seal recessed 9.65-9.90 mm (0.380-0.390 inch) into seal bore when measured from front of cover. A special tool (MFN 747B) and spacer (MFN 747C) are available to properly position the seal.

The timing gear cover is not doweled. Special tool MFN 747B can be used as a pilot when reinstalling. If tool is not available, reinstall crankshaft pulley to center the seal when reinstalling cover retaining screws. Assemble by reversing the disassembly procedure making sure pulley timing marks are aligned as shown in Fig. 27. Tighten the pulley retaining cap screw to a torque of 380-405 N·m (280-300 ft.-lbs.).

TIMING GEARS

All Gasoline Models

19. Refer to Fig. 28 for view of timing gear train with cover removed. Before attempting to remove any of the timing gears, first remove fuel tank, rocker arm cover and rocker arms to avoid the possibility of damage to pistons or valve train if camshaft or crankshaft should either one be turned independently of the other.

Number 1 piston will be at TDC on compression stroke and gears properly timed when timing marks are aligned as shown in Fig. 28. However, all timing marks will not be aligned each time piston is at TDC on compression stroke due to the odd number of teeth on idler gear. The crankshaft gear also drives the engine balancer, and balancer must be timed to crankshaft as outlined in paragraph 35. Governor drive gear does not need to be timed, but alignment of marks is a convenience for engine assembly.

Normal backlash between timing gears (except balancer idler gear) is 0.08-0.15 mm (0.003-0.006 inch). Backlash between balancer idler gear and its mating gears should be 0.15-0.23 mm (0.006-0.009 inch). Refer to the appropriate following paragraphs for renewal of gears, idler shaft or bushings if backlash is excessive or if parts are damaged.

20. IDLER GEAR AND HUB. Diametral clearance of idler gear on hub should be 0.058-0.119 mm (0.0023-0.0047 inch). Permissible end play 0.10-0.20 mm (0.004-0.008 inch). Idler gear bushing is renewable and must be reamed to 50.8 mm (2.000 inch) inside diameter after installation.

Idler gear hub is a light press fit in timing gear housing bore. Due to uneven spacing of mounting holes, hub can only be installed in one position. Tighten the three retaining cap screws to a torque of 27-33 N·m (20-24 ft.-lbs.). Measure end play with feeler gage (Fig. 29) after idler gear is installed.

21. CAMSHAFT GEAR. Camshaft gear is keyed to shaft and retained by a cap screw. Camshaft gear is a transition fit, 0.025 mm (0.001 inch) tight to 0.025 mm (0.001 inch) clearance, and threaded holes are provided for pulling gear.

Make sure timing marks on gear face outward, and tighten retaining cap screw to a torque of 61-68 N·m (45-50 ft.-lbs.) when installing gear.

22. GOVERNOR GEAR. The governor gear also drives the distributor. Gears contain timing marks which are convenient for assembly but not essential for engine timing, as timing can be accomplished externally when distributor is installed.

Governor gear is a slip fit on shaft. Shaft contains governor weight unit as well as distributor drive, and unit can be removed from rear as outlined in paragraph 28 after gear is off. When installing gear, tighten retaining nut to a torque of 31-34 N·m (23-25 ft.-lbs.).

23. CRANKSHAFT GEAR. Crankshaft gear is keyed to shaft and is a transition fit on shaft. It is usually possible to remove the gear using two small pry bars to move the gear forward. Engine balancer must be removed if puller is required.

24. TIMING THE GEARS. To install and time the gears, first install camshaft

Fig. 29—Idler gear end play can be checked with a feeler gage as shown. Specified end play is 0.10-0.20 mm (0.004-0.008 inch).

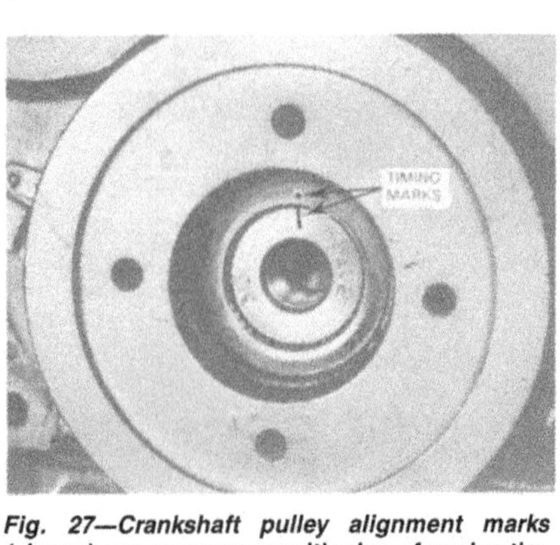

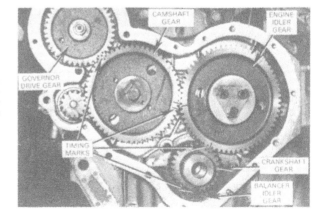

Fig. 27—Crankshaft pulley alignment marks (shown) ensure proper positioning of engine timing marks.

Fig. 28—Timing gear train with cover removed. Governor drive gear timing marks align only occasionally but alignment is not necessary for proper engine timing. Refer to text for procedure.

and crankshaft gears as outlined in the appropriate preceding paragraphs with timing marks to front. Be sure crankshaft gear is timed to balancer as outlined in paragraph 35. Turn the shafts until the appropriate timing marks point toward idler gear hub, then install idler gear with marks aligned as shown in Fig. 28. The governor gear timing marks may be aligned for convenience when all gears are removed, but alignment is not necessary for proper ignition timing. Secure idler gear as outlined in paragraph 20.

TIMING GEAR HOUSING

All Gasoline Models

25. To remove the timing gear housing, first remove timing gears as outlined in paragraphs 19 through 23 and the distributor and drive unit as in paragraph 28. Timing gear housing must be removed before camshaft can be withdrawn. Remove cap screws retaining timing gear housing to engine block and oil pan and lift off housing. Install by reversing the removal procedure.

CAMSHAFT

All Gasoline Models

26. To remove the camshaft, first remove timing gears, distributor drive unit (governor) and timing gear housing as outlined in paragraphs 19 through 25. Secure cam followers (tappets) in uppermost position and lift off thrust washer (Fig. 30), then withdraw camshaft from block bores.

Thrust washer retains camshaft and controls end play. Thrust washer thickness is 5.49-5.54 mm (0.216-0.218 inch); check the washer for correct thickness and for wear or scoring. Recommended camshaft end play is 0.10-0.41 mm (0.004-0.016 inch). Camshaft journal diameter and desired clearance are as follows:

Journal Diameter
Front . . . . .50.71-50.74 mm
(1.9965-1.9975 in.)
Center . . . . .50.46-50.48 mm
(1.9865-1.9875 in.)

Fig. 30—Camshaft is retained by thrust washer which can be removed only after removing timing gear housing.

Rear . . . . .49.95-49.97 mm
(1.9665-1.9675 in.)
Bearing
Clearance . . . . .0.064-0.140 mm
(0.0025-0.0055 in.)

CAM FOLLOWERS

All Gasoline Models

27. The mushroom type cam followers (tappets) operate directly in machined bores in engine block and can be removed from bottom of cylinder block after removing camshaft as outlined in paragraph 26 and oil pan. Keep cam followers in order as they are removed so they can be reinstalled in their original locations if reused. Always renew cam followers whenever camshaft is renewed.

The 18.987-19.012 mm (0.7475-0.7485 inch) diameter cam followers are available in standard size only. Diametral clearance of cam followers in block bores should be 0.038-0.095 mm (0.0015-0.0037 inch). Do not attempt to resurface face of cam followers if worn, pitted, scored or otherwise damaged.

DISTRIBUTOR DRIVE SHAFT

All Gasoline Models

28. Refer to Fig. 31 for exploded view. To remove or disassemble the unit, first

remove timing gear cover as outlined in paragraph 18 and governor gear as in paragraph 22. Remove the distributor as shown in Fig. 32 and governor shaft thrust plate. Disconnect governor and throttle linkage and unbolt and remove housings as shown in Fig. 33.

Unbolt and remove governor housing from drive body. Withdraw thrust bearing and distributor drive gear, then remove drive shaft assembly and governor weight unit.

Distributor drive shaft end play is 0.10-0.20 mm (0.004-0.008 inch), and thrust is forward because of governor action. Diametral clearance of shaft in housing bore is 0.03-0.08 mm (0.001-0.003 inch).

Fig. 32—Remove distributor clamp and lift out distributor assembly.

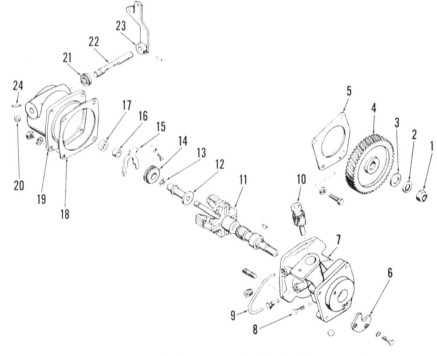

Fig. 31—Exploded view of governor and distributor drive assembly.

1. Nut	8. Distributor drive	12. Sleeve	18. Gasket
2. Lockwasher	gear retaining pin	13. Snap ring	19. Housing
3. Plain washer	9. Oil line	14. Thrust bearing	20. Plug
4. Drive gear	10. Distributor drive	15. Yoke	21. Seal
5. Gasket	gear	16. Bushing	22. Shaft
6. Thrust plate	11. Drive shaft assy.	17. Seal	23. Lever
7. Drive body			24. Shaft retaining pin

Fig. 33—Removing governor housing assembly.

Assemble by reversing the disassembly procedure. Turn crankshaft until No. 1 piston is at TDC on compression stroke and timing mark on crankshaft pulley is aligned as shown in Fig. 24. Then install distributor gear with drive slot perpendicular to crankshaft and offset to rear as shown in Fig. 34. Install distributor and turn body until points just begin to open, then install and tighten clamp. Adjust governed speed as outlined in paragraph 112 and timing as in paragraph 135.

ROD AND PISTON UNITS

All Gasoline Models

29. Connecting rod and piston units are removed from above after removing cylinder head, oil pan, engine balancer and rod bearing caps. Cylinder numbers are stamped on connecting rod and cap. Make certain correlation numbers are in register and face away from camshaft side of engine when reassembling. Tighten connecting rod nuts to a torque of 102-108 N·m (75-80 ft.-lbs.).

PISTONS, SLEEVES AND RINGS

All Gasoline Models

30. The aluminum alloy pistons have a combustion chamber cavity cast into piston crown. New pistons are alike and nondirectional but if old pistons are used, pistons should be installed in same cylinder and in same direction as originally installed. Pistons are available in standard size only. Heating piston to 40°-50° C (100°-120° F) in hot water or oil will make installation of piston pin easier.

Each piston is fitted with a plain faced chrome top ring which may be installed either side up as may the plain faced second compression ring. Refer to appropriate Fig. 35 or Fig. 36 for correct installation of rings in grooves. The third ring is internally stepped and should be installed in groove with grooves of inner diameter toward top. The fourth, oil

control ring for MF255 is chrome plated, spring loaded and can be installed either side up. The fourth, oil control ring for MF265 is three piece with chrome plated rails. Refer to the following for recommended ring end gap and side clearance in groove.

Ring End Gap:
Top . . . . 0.41-0.51 mm
                                (0.016-0.020 inch)
Second and Third . . . . . 0.31-0.43 mm
                                (0.012-0.017 inch)
Fourth, MF255 . . . . 0.41-0.53 mm
                                (0.016-0.021 inch)
Fourth, MF265 . . . . 0.38-1.40 mm
                                (0.015-0.055 inch)

Ring Side Clearance:
Top . . . . 0.051-0.102 mm
                                (0.002-0.004 inch)
Second and
    Third . . . . 0.048-0.099 mm
                                (0.0019-0.0039 inch)
Fourth, MF255 . . . . . . 0.064-0.114 mm
                                (0.0025-0.0045 inch)
Fourth, MF265 . . . . Not Applicable

Production sleeves are tight press a fit in cylinder block and are finished after installation. Service sleeves are a transition fit and are prefinished. Sleeves should not be bored and oversize pistons are not available. When installing new sleeves, make sure sleeves and bores are

Fig. 34—With No. 1 piston at TDC on compression stroke, distributor drive slot should be nearly perpendicular to crankshaft and offset to rear as shown.

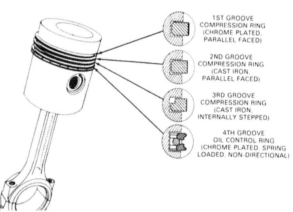

Fig. 35—Cross-sectional view showing correct installation of piston rings in piston grooves on Model MF255.

1ST GROOVE
COMPRESSION RING
(CHROME PLATED,
PARALLEL FACED)

2ND GROOVE
COMPRESSION RING
(CAST IRON,
PARALLEL FACED)

3RD GROOVE
COMPRESSION RING
(CAST IRON,
INTERNALLY STEPPED)

4TH GROOVE
OIL CONTROL RING
(CHROME PLATED, SPRING
LOADED, NON-DIRECTIONAL)

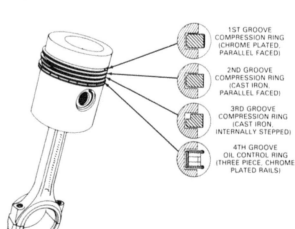

Fig. 36—Cross-sectional view showing correct installation of piston rings in piston grooves on Model MF265.

1ST GROOVE
COMPRESSION RING
(CHROME PLATED,
PARALLEL FACED)

2ND GROOVE
COMPRESSION RING
(CAST IRON,
PARALLEL FACED)

3RD GROOVE
COMPRESSION RING
(CAST IRON,
INTERNALLY STEPPED)

4TH GROOVE
OIL CONTROL RING
(THREE PIECE, CHROME
PLATED RAILS)

absolutely clean and dry, then chill the sleeves and press fully into place by hand. Top edge of sleeve should extend 0.762-0.889 mm (0.030-0.035 inch) above gasket face of cylinder block. Inside diameter of installed sleeve should be 98.476-98.501 mm (3.877-3.878 inches).

PISTON PINS

All Gasoline Models

31. The full floating piston pins are retained in piston bosses by snap rings and are available in standard size only. The renewable connecting rod bushing must be final sized after installation to 25.41143-25.43048 mm (1.00045-1.00120 inches). Be sure the predrilled oil hole in bushing is properly aligned with hole in top of connecting rod and install bushing from chamfered side of bore.

Piston pin diameter is 25.395-25.400 mm (0.9998-1.0000 inch). Piston pin should have 0.0114-0.0355 mm (0.00045-0.00140 inch) clearance in rod bushing and 0.0-0.010 mm (0.0-0.0004 inch) clearance in piston.

NOTE: Heating piston to 40°-50° C (100°-120° F) in hot water or oil will make installation of piston pin in piston easier.

CONNECTING RODS AND BEARINGS

All Gasoline Models

32. Connecting rod bearings are precision type, renewable from below after removing oil pan, balancer unit and rod bearing caps. When renewing bearing shells, be sure the projection engages the milled slot in rod and cap and that correlation marks are in register and face away from camshaft side of engine.

Connecting rod bearings should have a diametral clearance of 0.038-0.076 mm (0.0015-0.0030 inch) on the 63.475-63.487 mm (2.4990-2.4995 inch) diameter crankpin. Recommended connecting rod side clearance is 0.216-0.368 mm (0.0085-0.0145 inch). Renew the self-locking connecting rod nuts and tighten to a torque of 102-108 N·m (75-80 ft.-lbs.).

CRANKSHAFT AND BEARINGS

All Gasoline Models

33. The crankshaft is supported in five precision type main bearings. To remove the rear main bearing cap, it is first necessary to remove engine, clutch, flywheel and rear oil seal. All other main bearing caps can be removed after removing oil pan and engine balancer.

Upper and lower main bearing inserts are not interchangeable. The upper (block) half is slotted to provide pressure lubrication to crankshaft and connecting rods. Inserts are interchangeable in pairs for all journals except center main bearing. The center journal controls crankshaft end thrust and renewable thrust washers are installed at front and rear of cap and block bearing bore. Lower half of insert is anchored by a tab to bearing cap; upper half can be rolled out and renewed after cap is removed.

Main bearing inserts are available in undersizes of 0.25, 0.50 and 0.75 mm (0.010, 0.020 and 0.030 inch) as well as standard. Thrust washers are available in standard size and 0.19 mm (0.0075 inch) oversize. Oversize thrust washers may be installed on one side in combination with standard washers on the other side for an adjustment of 0.19 mm (0.0075 inch), or they can be installed on both sides for an adjustment of 0.38 mm (0.015 inch). Recommended crankshaft end play is 0.05-0.38 mm (0.002-0.015 inch).

Note that main bearing caps are not interchangeable. Caps are numbered 1 through 5 from front to rear and are also marked with a cylinder block identification number that must match number stamped on bottom of cylinder block. Recommended tightening torque for main bearing cap screws is 230-244 N·m (170-180 ft.-lbs.). Refer to paragraph 37 for installation procedure for rear seal

Fig. 37—Crankshaft end play should be within limits of 0.05-0.38 mm (0.002-0.015 Inch).

Fig. 38—Installed view of engine balancer with timing marks aligned. Refer to text for installation procedure.

and oil pan bridge piece. Refer to the following specifications when checking crankshaft.

Main Journal
Standard Diameter . 76.162-76.175 mm
(2.9985-2.9990 in.)
Maximum Allowable Out-
of-Round or Taper . . . . . . 0.025 mm
(0.001 in.)
Bearing Diametral
Clearance . . . . 0.064-0.114 mm
(0.0025-0.0045 in.)
**Center Main
Journal Width**
Standard . . . . 44.145-44.221 mm
(1.738-1.741 in.)
Maximum Allowable . . . . . 44.678 mm
(1.759 in.)
Crankpin
Standard Diameter 63.475-63.487 mm
(2.4990-2.4995 in.)
Bearing Diametral
Clearance 0.038-0.076 mm
(0.0015-0.0030 in.)
**Crankshaft End
Play** . . . . 0.05-0.38 mm
(0.002-0.015 in.)

ENGINE BALANCER

All Gasoline Models

34. OPERATION. The Lanchester type engine balancer consists of two unbalanced shafts which rotate in opposite directions at twice crankshaft speed. The inertia of the shaft weights is timed to cancel out natural engine vibration, thus producing a smoother running engine. The balancer is correctly timed when the balance weights are at their lowest point when pistons are at TDC and BDC of their stroke.

The balancer unit is driven by the crankshaft timing gear through an idler gear attached to balancer frame. The engine oil pump is mounted at rear of balancer frame and driven by the balancer shaft. Refer to Figs. 38 through 40.

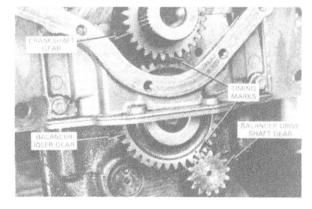

35. REMOVE AND REINSTALL. The balancer assembly can be removed after removing the oil pan and mounting cap screws. Engine oil is pressure fed through balancer frame and cylinder block. Balancer frame bearings are also pressure fed.

When installing balancer with engine in tractor, timing marks will be difficult to observe without removing gear cover. The balancer assembly can be safely installed as follows: Turn crankshaft until No. 1 and No. 4 pistons are at the exact bottom of their stroke. Remove balancer idler gear (5—Fig. 40), if necessary, and reinstall with single punch-marked tooth of idler gear meshed between the two marked teeth on weight drive shaft as shown in Fig. 38. Install balancer frame with balance weights hanging normally. If carefully installed, timing will be correct although timing marks at front of engine may not be aligned. (Marks align only occasionally when engine is running.) If engine is mounted in stand or tractor front end is removed, timing marks can be observed by removing timing gear cover.

NOTE: Balancer can be safely installed with No. 1 and 4 pistons at either TDC or BDC. BDC is selected because interference between connecting rod and balance weights can give warning if unit is badly out of time.

With balancer correctly installed, tighten the retaining cap screws to a torque of 43-49 N·m (32-36 ft.-lbs.) and complete the assembly by reversing the removal procedure.

36. OVERHAUL. Refer to Fig. 40 for an exploded view of balancer frame and associated parts. To disassemble the removed balancer unit, unbolt and remove oil pump housing (20) and associated parts, and idler gear (5) and associated parts. Set screws (S) retaining balance weights (15 and 16) are installed using "Loctite 242

Threadlocker." Loosen screws, then push balance shafts (11 and 12) forward out of frame and weights.

NOTE: Use care when removing shafts not to allow keys (13) to damage frame bushings as bushings are not available as a service item.

Inspect all parts for wear or damage. Bushings in balancer frame are not available separately. If bushings are excessively worn, balancer frame must be renewed. Refer to Fig. 40 for identification of parts and following table for specifications.

Drive Shaft (11)

Front Journal
  Diameter . . . . . . .31.725-31.737 mm
  (1.2490-1.2495 in.)
Clearance in Front
  Bushing . . . . .0.051-0.106 mm
  (0.0020-0.0042 in.)
Rear Journal
  Diameter . . . . . . .25.375-25.387 mm
  (0.9990-0.9995 in.)
Clearance in Rear
  Bushing . . . . .0.046-0.089 mm
  (0.0018-0.0035 in.)

Driven Shaft (12)

Front Journal
  Diameter . . . . . . .31.704-31.724 mm
  (1.2482-1.2490 in.)
Clearance in Front
  Bushing . . . . .0.064-0.122 mm
  (0.0025-0.0048 in.)
Rear Journal
  Diameter . . . . . . .25.367-25.379 mm
  (0.9987-0.9992 in.)
Clearance in Rear
  Bushing . . . . .0.053-0.096 mm
  (0.0021-0.0038 in.)

Idler Gear (5)

Clearance on Hub . . .0.025-0.081 mm
  (0.0010-0.0032 in.)
End Play . . . . .0.20-0.35 mm
  (0.008-0.014 in.)

When assembling balancer, use Loctite 242 Threadlocker when installing screws retaining gears (14—Fig. 40) to balance weights (15 and 16) and the set screws (S) retaining balance weights to shafts. Make sure flat surfaces of weights are aligned when installed as shown in Fig. 39.

After renewing balancer assembly or any parts of balancer, engine should be operated at idle speed for ½ hour,

Fig. 39—Assembled view of removed engine balancer and oil pump unit. Refer to Fig. 40 for exploded view.

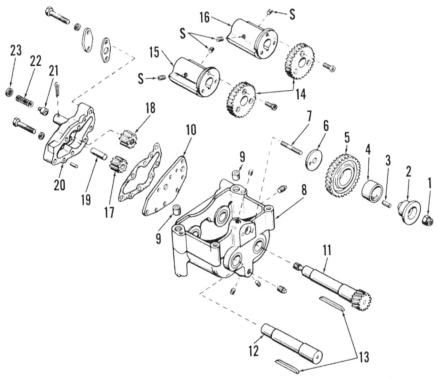

Fig. 40—Exploded view of Lanchester type engine balancer, engine oil pump and associated parts.

1. Locknut	7. Stud	13. Key	19. Shaft
2. Hub	8. Frame	14. Gear	20. Pump body
3. Dowel	9. Ring dowels	15. Balance weight	21. Valve piston
4. Bushing	10. Plate	16. Balance weight	22. Valve spring
5. Idler gear	11. Drive shaft	17. Pump gear	23. Cap
6. Washer	12. Driven shaft	18. Pump gear	S. Set screw

followed by another ½ hour at 1000 rpm to properly run-in new parts before returning tractor to normal service.

REAR OIL SEAL

All Gasoline Models

37. The asbestos rope type rear oil seal is contained in a two-piece seal retainer attached to rear face of engine block as shown in Fig. 41. The seal retainer can be removed after removing flywheel.

The rope type crankshaft seal is precision cut to length, and must be installed in retainer halves with 0.25-0.50 mm (0.010-0.020 inch) of seal end protruding from each end of retainer groove. Do not trim the seal. To install the seal, clamp each retainer half in a vise as shown in Fig. 42. Make sure seal groove is clean. Start each end in groove with the specified amount of seal protruding. Allow seal rope to buckle in the center until about 25 mm (1 inch) of each end is imbedded in groove, work center of seal into position, then roll with round bar as shown. Repeat the process with other half of seal retainer.

Fig. 41—Rear view of engine block showing oil seal retainer installed.

Fig. 42—Use a round bar to bed the asbestos rope seal in retainer half. Refer to text for details.

When installing cylinder block bridge piece, insert end seals as shown in Fig. 43. Use a straightedge as shown in Fig. 44 to make sure bridge piece is flush with rear face of cylinder block.

Coat both sides of retainer gasket and end joints of retainer halves with a suitable gasket cement. Coat surface of rope seal with graphite grease. Install retainer halves and cap screws loosely and tighten clamp screws thoroughly before tightening the retaining cap screws.

FLYWHEEL

All Gasoline Models

38. To remove the flywheel, first separate engine from transmission housing as outlined in paragraph 148 and remove the clutch. Flywheel is secured to crankshaft flange. To properly time the flywheel, align the seventh (unused) hole in flywheel with untapped hole in crankshaft flange. Tighten flywheel retaining cap screws to a torque of 100-108 N•m (74-80 ft.-lbs.).

OIL PAN

All Gasoline Models

39. The heavy cast iron oil pan serves as part of tractor frame and as attaching point for tractor front support. To remove the oil pan, first drain the oil.

Fig. 43—Cylinder block bridge is equipped with end seals as shown.

Support tractor under transmission housing, remove cap screws securing oil pan to cylinder block, front support and transmission housing and lower the oil pan from cylinder block.

Install by reversing the removal procedure.

OIL PUMP

All Gasoline Models

40. The gear type oil pump is mounted on engine balancer frame and driven by balancer shaft. Refer to Fig. 40. Oil pump can be removed after removing engine oil pan.

The thickness of gears (17 and 18) should be 25.35-25.40 mm (0.998-1.000 inch). End clearance of gears with a 0.10 mm (0.004 inch) gasket installed should be 0.08-0.20 mm (0.003-0.008 inch). Radial clearance of gears in pump body should be 0.05-0.23 mm (0.002-0.009 inch). Examine gears, body and end plate for wear or scoring and renew any parts which are questionable.

RELIEF VALVE

All Gasoline Models

41. The plunger type relief valve is located in oil pump body as shown in Fig. 40. Oil pressure should be 210-380 kPa (30-55 psi) at full engine speed with engine at normal operating temperature. Valve spring is retained by a cap and cotter pin.

Fig. 44—Use a straightedge to align the cylinder block bridge.

PERKINS AD4.203 DIESEL ENGINE AND COMPONENTS

Early MF255 diesel models are equipped with a Perkins AD4.203 four cylinder diesel engine. The engine has a bore of 91.49 mm (3.6 inches), a stroke of 127 mm (5 inches) and displacement of 3.3 L (203 cu. in.).

R&R ENGINE ASSEMBLY

All Models So Equipped

42. To remove the engine and clutch as a unit, first drain cooling system, and if engine is to be disassembled, drain oil pan. Remove front grille panels, disconnect battery cables and remove hood. Disconnect headlight wires, radiator hoses, air cleaner hose and hydraulic cooler lines. Detach battery cable from starter motor, remove left rear side panel and disconnect power steering oil lines. Support tractor with a jack at front of transmission and place wedges between front axle and axle support to prevent tipping. Remove alternator to gain access to front end mounting bolt, then support front axle so that it can be moved forward when detached. Unbolt and separate front assembly from engine.

Shut fuel off at tank, disconnect fuel gage sending unit wire, main fuel line and fuel return line, then remove tank and support plate. Disconnect main wiring harness from starter motor and water temperature gage sending unit. Disconnect oil pressure gage line, tachometer drive cable, fuel shut-off cable, throttle rod and ether starting aid line (if so equipped). Disconnect fuel line between fuel filter and injection pump at injection pump end. Disconnect fuel return line between fuel filter and injector leak-off manifold at leak-off manifold end. Disconnect exhaust pipe from models with low exhaust. On all models, remove fuel line from between fuel lift pump and fuel filter. Be sure that all wires, linkage and tubes are positioned out of the way for engine removal.

Attach hoist to lifting eyes of engine then remove starting motor and screws attaching engine to transmission housing. Guide studs should be used while separating and rejoining engine to transmission.

Reassemble in reverse of removal procedure. Coat input shaft splines lightly with lithium base grease before assembling. Be sure that spacers (Fig. 45) are correctly positioned when reinstalling front end.

CYLINDER HEAD

All Models So Equipped

43. REMOVE AND REINSTALL. To remove the cylinder head, drain cooling system and remove hood. Disconnect fuel gage sending unit wire, main fuel line and fuel return line then remove fuel tank, heat shield and support plate. Disconnect upper radiator hose, air cleaner hose, water temperature gage sending unit wire, breather tube and

coolant bypass tube. Remove the oil line (Fig. 47), exhaust manifold, main fuel line (Fig. 46), fuel leak-off line, fuel leak-off manifold, high pressure fuel lines, injectors, rocker arm cover and rocker arm shaft assembly. Loosen cylinder head stud nuts in reverse of sequence shown in Fig. 48.

Check cylinder head gasket surface for distortion using a straightedge and feeler gage. If cylinder head is warped in excess of 0.10 mm (0.004 inch), head should be resurfaced. A maximum of 0.30 mm (0.012 inch) of material may be removed to true the gasket surface, providing overall thickness of cylinder head is not reduced below 75.69 mm (2.980 inches) and injector nozzle protrusion does not exceed 4.6 mm (0.181 inch). Valve head recession must also be maintained within specified limits as outlined in paragraph 44. Original cylinder head thickness is 75.82-76.58 mm (2.985-3.015 inches).

Clean mating surfaces of block and head, then position head gasket with "TOP FRONT" marking facing up and toward the front. Install cylinder head and tighten fasteners in sequence shown in Fig. 48 in 15-20 N·m (10-15 ft.-lbs.) increments until final torque of 95-102 N·m (70-75 ft.-lbs.) is obtained.

Turn tappet adjusting screws several turns clockwise into tappets to make sure valves do not strike the pistons when rocker arms are installed. Install rocker shaft assembly and tighten retaining nuts to 28-33 N·m (21-24 ft.-lbs.) torque. Adjust valve clearance cold to 0.30 mm (0.012 inch) for intake and exhaust valves. Refer to paragraph 49 for adjustment procedure. Remainder of installation is reverse of removal procedure.

Retorque cylinder head and adjust valve clearance after engine has reached operating temperature. Valve clearance should be 0.25 mm (0.010 inch) for all valves after engine is at operating temperature.

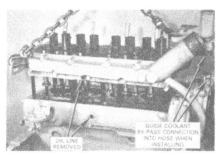

Fig. 47—View showing cylinder head being lifted off. Notice location of coolant bypass hose.

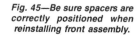
Fig. 45—Be sure spacers are correctly positioned when reinstalling front assembly.

Fig. 46—View of MF255 diesel showing temperature gage sending unit, fuel leak-off line, leak-off manifold and main fuel line.

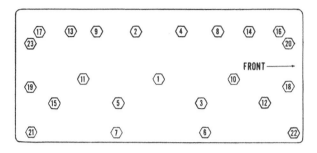

Fig. 48—Tighten cylinder head stud nuts evenly to final torque of 95-102 N·m (70-75 ft.-lbs.) using sequence shown.

VALVES AND SEATS

All Models So Equipped

44. Intake and exhaust valves seat directly in the cylinder head. Valve heads and seat locations are numbered consecutively from front to rear. Any replacement valves should be so marked prior to installation.

Intake and exhaust valves have a face angle of 45°. Valve seat angle for production is 46°; however, 45° can be used for service. Desired seat width is 1.6-2.4 mm (1/16 to 3/32 inch).

Valve heads should be recessed a specified amount into the cylinder head. The recession (clearance) between gasket surface of cylinder head and top of valve head can be measured using a straightedge and feeler gage as shown in Fig. 49. During production, recession is held within the limits of 1.52-1.85 mm (0.060-0.073 inch) for intake valves and 1.55-1.88 mm (0.061-0.074 inch) for exhaust valves. A maximum recession of 2.13 mm (0.084 inch) is permissible before renewing valves or cylinder head.

Service valve seat inserts are available for exhaust valves. Cylinder head must be machined to provide appropriate size counterbore for installation of valve seat inserts. See Fig. 50 for specified dimensions of counterbore. It is recommended that new valve guides be installed prior to machining cylinder head to provide an accurate pilot for cutting

tool. Reface valve seat to 45 degrees after installation, then check for specified valve head recession.

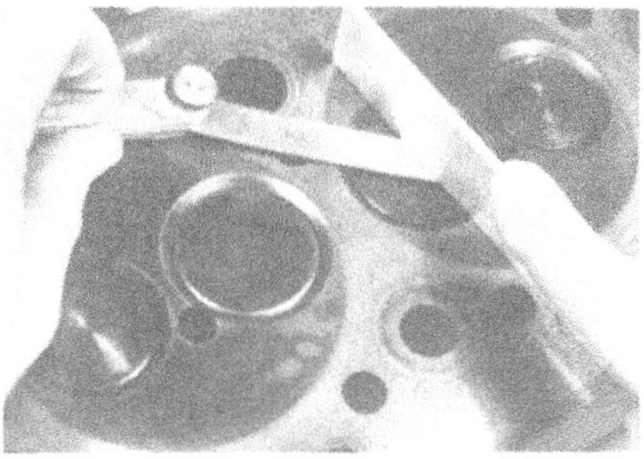

VALVE GUIDES

All Models So Equipped

45. The intake and exhaust valve guides are interchangeable and should be renewed if clearance between new valve and guide exceeds 0.114 mm (0.0045 inch). Press guides out and in from top only. Install guides with 20° chamfer down (in port) and 45 degrees chamfer up as shown in Fig. 51. Top of guide should extend 14.73-15.09 mm (0.580-0.594 inch) from top surface of cylinder head.

Valve guides are presized and do not require reaming when properly installed. Inside diameter of new guide (installed) should be 7.998-8.014 mm (0.3145-0.3155 inch). Valve stem diameter should be 7.899-7.925 mm (0.311-0.312 inch). Be sure to reseat valves after renewing valve guides.

VALVE SPRINGS

All Models So Equipped

46. Springs, retainers and locks are interchangeable for intake and exhaust valves. Springs are nondirectional and may be installed either end up. Renew

the springs if they are distorted, discolored or fail to meet the following test specifications: When compressed to a length of 38.1 mm (1.5 inches), test load should be 93-110 kg (20-3/4 to 24-3/4 pounds).

CAM FOLLOWERS

All Models So Equipped

47. The mushroom type cam followers (tappets) operate directly in machined bores in cylinder head. The cam followers actuate directly against camshaft and rocker arms without the use of push rods.

To remove the cam followers after cylinder head is off, first remove the adjusting screw and locknut, then withdraw the cam follower from its bore. Keep cam followers in order of removal so they can be reinstalled in original locations.

The 15.805-15.843 mm (0.62225-0.62375 inch) diameter cam followers are available in standard size only. Diametral clearance of cam followers in cylinder head bores should be 0.019-0.089 mm (0.00075-0.00350 inch). Cam followers should be renewed whenever camshaft is renewed.

ROCKER ARMS

All Models So Equipped

48. The rocker arms and shaft assembly can be removed after removing the hood, fuel tank and rocker arm cover. The rocker arms are right-hand and left-hand units and should be installed on shaft as shown in Fig. 52. Desired diametral clearance between new rocker arms and new shaft is 0.0190-0.0889 mm (0.00075-0.00350 inch). Renew shaft and/or rocker arm if clearance is excessive.

The amount of oil circulating to the rocker arms is regulated by the rotational position of the rocker shaft in the

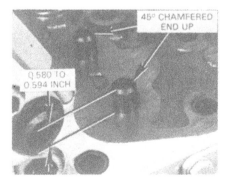

Fig. 49—Valve heads should be recessed specified amount below cylinder head gasket surface. Measure as shown using straightedge and feeler gage.

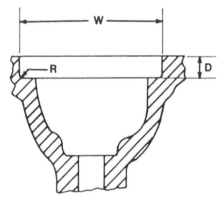

Fig. 50—A counterbore must be machined in cylinder head for installation of exhaust valve seat insert.

D. 7.87-7.92 mm
(0.310-0.312 in.)
R. 0.38 mm radius
(0.015 in.)

W. 42.62-42.64 mm
(1.678-1.679 in.)

Fig. 51—Valve guides should be installed to the correct height with 45 degree chamfered end up as shown.

support brackets. This position is indicated by a slot in one end of the rocker shaft as shown in Fig. 53. When the slot is positioned horizontally, the maximum oil circulation is obtained. In production, the slot is positioned 30° from the vertical and the position indicated by a punch mark (P) on the adjacent support bracket. When reassembling, position the rocker shaft slot as indicated by the punch mark and check the assembly for proper lubrication. The shaft will not normally need to be moved from the marked position. The slot must be located at rear of shaft as indicated in Fig. 52.

VALVE CLEARANCE

All Models So Equipped

49. The recommended cold valve clearance (tappet gap) setting is 0.30 mm (0.012 inch) for both the intake and exhaust valves. Static setting of all valves can be made from just two crankshaft positions using the procedure outlined in Figs. 54 and 55.

Remove timing plug from left side of flywheel adapter housing and turn crankshaft until the "TDC 1" timing mark is aligned with pointer as shown in Fig. 56.

Check the rocker arms for the front and rear cylinders. If rear rocker arms are tight and front rocker arms loose, No. 1 piston is on the compression stroke; adjust the four valves shown in

Fig. 54. If front rocker arms are tight and rear rocker arms loose, No. 4 piston is on the compression stroke; adjust the four valves shown in Fig. 55. Turn crankshaft one complete turn until "TDC 1" timing mark is again aligned, then adjust the remaining valves.

Recheck adjustment after engine has reached operating temperature. Clearance hot should be 0.25 mm (0.010 inch) for all valves.

VALVE TIMING

All Models So Equipped

50. To check the valve timing when engine is assembled, first turn crankshaft until No. 1 piston is at TDC

on compression stroke. Adjust the clearance of front (intake) valve to 1.14 mm (0.045 inch). Insert a 0.05 mm (0.002 inch) feeler gage between rocker arm and front valve, then slowly turn crankshaft in normal direction of rotation (clockwise viewed from front) until feeler gage just becomes tight. At this point, the "TDC 1" mark on flywheel should again be aligned with timing pointer as shown in Fig. 56.

NOTE: Timing may be considered correct if "TDC 1" mark is within 8 mm (5/16 inch) of alignment with timing pointer. Readjust the front valve to the recommended cold or hot clearance setting before attempting to start the engine.

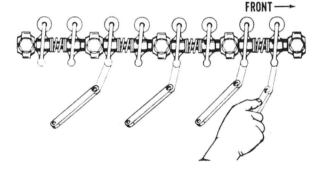

Fig. 54—With "TDC 1" timing mark aligned as shown in Fig. 56 and No. 1 piston on compression stroke, adjust the indicated valves. With adjustment completed, turn crankshaft one revolution and refer to Fig. 55.

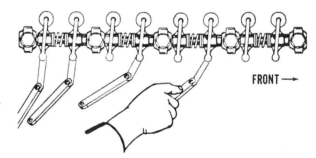

Fig. 55—With "TDC 1" timing mark aligned as shown in Fig. 56 and No. 4 piston on compression stroke, adjust the indicated valves. With adjustment completed, turn crankshaft one revolution and refer to Fig. 54.

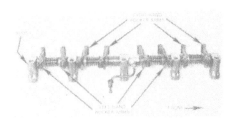

Fig. 52—View of correctly assembled rocker arm assembly. Slot in end of shaft should be toward rear and at angle shown in Fig. 53.

Fig. 53—Slot in end of rocker arm shaft should be set at angle shown to meter the correct amount of oil.

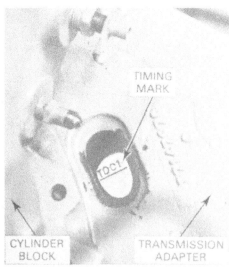

Fig. 56—View of "TDC 1" timing mark aligned for valve adjustment as outlined in paragraph 49.

There is no adjustment for valve timing. If timing is incorrect, the timing gears are not meshed correctly. Timing gear marks (Fig. 59) may be viewed after timing cover is removed as outlined in paragraph 51. Refer to paragraph 52 for explanation of timing marks.

TIMING GEAR COVER

All Models So Equipped

51. To remove the timing gear cover first drain cooling system and engine oil. Remove front grille panels, disconnect battery cables and remove hood. Disconnect headlight wires, radiator hoses, air cleaner hose and hydraulic cooler lines. Detach battery cable from starter motor, remove left rear side panel and disconnect power steering oil line. Support tractor with jack at front of transmission and place wedges between front axle and axle support to prevent tipping. Remove alternator to gain access to front end mounting bolt, then support front axle so it can be rolled forward when detached. Unbolt and separate front assembly from engine.

Remove water pump, fan belt and crankshaft pulley. Unbolt and remove lower timing gear cover, then unbolt and remove upper cover.

The timing gear cover is not located by dowels or similar fixture and must be correctly positioned before tightening the retaining screws. Special tool (MFN 747AA) is available to center timing gear cover around crankshaft before seal is installed. The special tool can also be used to press the new seal into the installed cover so seal is 8.13-8.38 mm (0.32-0.33 inch) from front face of cover. Tighten crankshaft pulley retaining screw to 136-149 N·m (100-110 ft.-lbs.) torque.

TIMING GEARS

All Models So Equipped

52. Timing gears may be viewed as shown in Fig. 59 after cover is removed as described in paragraph 51. Due to the odd number of teeth on the two idler gears, all timing marks will align only once about every 2000 revolutions. Timing is therefore more easily checked by removing one or both idler gears.

Before attempting to remove any of the timing gears, first remove fuel tank, rocker arm cover and rocker arms assembly to avoid the possibility of damage to pistons or valve train if camshaft or crankshaft should be turned with part of gears removed.

Recommended timing gear backlash is 0.08-0.15 mm (0.003-0.006 inch) between either of the two idler gears and any of their mating gears in the timing gear train. Backlash between oil pump idler gear and crankshaft or oil pump gear should be 0.15-0.23 mm (0.006-0.009 inch). Replacement gears are available in standard size only. If backlash is not within the specified limits, renew idler gears, idler gear shafts and/or the other gears concerned.

Refer to the appropriate following paragraphs for removal, installation and timing of the gears:

53. IDLER GEARS AND HUBS. Idler gears should have a diametral clearance of 0.058-0.119 mm (0.0023-0.0047 inch) on idler gear hubs. Idler hub retainer plates and the flanged bushings in the lower idler gear (4—Fig. 58) are renewable. End play of assembled idler gear should be 0.10-0.20 mm (0.004-0.008 inch). Hubs (H—Fig. 61) are interchangeable for the two idler gears.

Fig. 57—View showing adjustment of valve clearance (tappet gap).

Retaining stud holes permit hubs to be installed only one way, with oil holes (B) aligned. Idler hubs are a light press fit in engine block and can be pried out if renewal is indicated. Refer to paragraph 57 for installation and timing of idler gears.

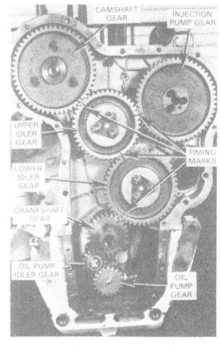

Fig. 59—View of timing gears with marks correctly aligned. Because of the number of teeth on idler gears, all marks will align only about once every 2000 crankshaft revolutions.

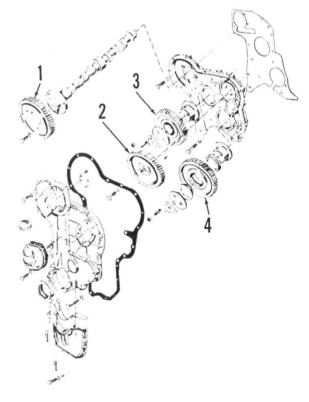

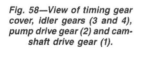
Fig. 58—View of timing gear cover, idler gears (3 and 4), pump drive gear (2) and camshaft drive gear (1).

Fig. 60—View of idler gear (3) shown also in Fig. 58.

properly aligned, reinstall idler gear retainer washers, locks and stud nuts.

TIMING GEAR HOUSING

All Models So Equipped

58. To remove the timing gear housing, first remove timing gears, engine oil pan, fuel injection pump and power steering pump. Block up the cam followers and withdraw camshaft, then unbolt and remove timing gear housing. A small, cup-type expansion plug which closes front end of oil gallery is accessible after removing the housing.

Install by reversing the removal procedure.

CAMSHAFT

All Models So Equipped

54. CAMSHAFT GEAR. The cap screw holes in camshaft and camshaft gear (1—Fig. 58) are evenly spaced and the gear can be installed in three positions, only one of which is correct. The camshaft gear is correctly installed when the stamped "D" on camshaft gear is aligned with the stamped "D" on camshaft mounting flange as shown in Fig. 62.

55. INJECTION PUMP DRIVE GEAR. The injection pump drive gear (2—Fig. 58) is retained to the pump adapter by three cap screws. When installing the gear, align dowel pin (Fig. 63) with the slot in pump adapter (P), then install the retaining cap screws. The injection pump drive gear and adapter are supported by the injection pump rotor bearings.

56. CRANKSHAFT GEAR. The crankshaft timing gear is keyed to the shaft and is a transition fit, 0.025 mm (0.001 inch) tight to 0.025 mm (0.001 inch) loose, on shaft. It is usually possible to remove the gear using two small

pry bars to move the gear forward. Use a suitable puller and adapter if gear cannot be removed with pry bars.

57. TIMING THE GEARS. To install and time the gears, first make sure camshaft gear, injection pump drive gear and crankshaft gear are correctly installed as outlined in paragraphs 54, 55 and 56. Turn crankshaft until key is pointing up as shown in Fig. 59, and install lower idler with one timing mark aligned with mark on crankshaft gear and the other mark pointing toward upper idler gear hub. Turn camshaft gear and pump drive gear so timing marks point toward upper idler gear hub, then install upper idler gear with all timing marks aligned. When timing marks are

59. To remove the camshaft, first remove timing gear cover as outlined in paragraph 51, then remove fuel tank, rocker arm cover and rocker arm assembly. Secure the valve tappets (cam followers) in the uppermost position, remove the fuel lift pump, then withdraw camshaft and gear as a unit as shown in Fig. 65.

Camshaft end play is controlled by the thrust plate (Fig. 67) attached to the timing gear cover and the thrust washer (A—Fig. 66) behind camshaft timing

Fig. 62—Retaining cap screw holes are evenly spaced in camshaft and camshaft timing gear. Gear is correctly installed when the stamped timing letter is aligned with similar mark on camshaft mounting flange.

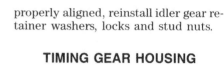

Fig. 61—Oil feed passage (B) in cylinder block and idler gear hub (H) must be aligned when hub is installed.

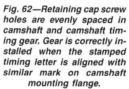

Fig. 63—Dowel in pump drive gear fits in milled slot in injection pump drive shaft (P) for correct pump timing. Timing marks in the entire timing gear train must be aligned as shown in Fig. 59 when gear is installed.

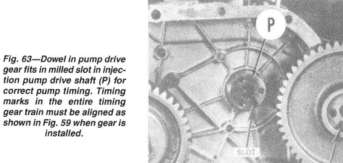

gear. Thrust washer (A) is 3.18-3.30 mm (0.125-0.130 inch) thick when new.

The camshaft runs in three journal bores machined directly in engine block. The front and rear camshaft bearings are gravity lubricated by return oil from the rocker arms. The center journal is pressure lubricated by an external oil line. The center journal, in turn, meters oil to the rocker shaft and cylinder head through a second short oil feed line.

Camshaft bearing journals have a recommended diametral clearance of 0.10-0.20 mm (0.004-0.008 inch) in all three bearing bores. Journal diameters are as follows:

Journal Diameter
Front . . . . .47.473-47.498 mm
(1.869-1.870 in.)
Center . . . . .47.219-47.244 mm
(1.859-1.860 in.)
Rear . . . . .46.711-46.736 mm
(1.839-1.840 in.)

ROD AND PISTON UNITS

All Models So Equipped

60. Connecting rod and piston units are removed from above after removing cylinder head, oil pan and rod bearing caps. Be sure to remove carbon and ring wear ridge (if present) from top of cylinders before pushing pistons out of cylinders.

Fig. 65—Camshaft and timing gear may be withdrawn as a unit as shown, refer to text.

Fig. 66—Camshaft thrust washer (A) is position-ed in engine block by locating pin (B).

Cylinder numbers are stamped on the connecting rod and cap. When reinstalling rod and piston units, make certain the correlation numbers are in register and face away from camshaft side of engine. Refer to Fig. 68. Tighten the connecting rod nuts to a torque of 54-61 N·m (40-45 ft.-lbs.).

PISTONS, SLEEVES AND RINGS

All Models So Equipped

61. The aluminum alloy, cam ground pistons are supplied in standard size only and are available in a kit consisting of piston, pin and rings for one cylinder. The toroidal combustion chamber is off-set in piston crown and piston is marked (A—Fig. 68) for correct assembly. Be sure that connecting rod is assembled to piston with correlation marks (C) toward left side. Heating piston to 40°-50° C (100°-120° F) in hot water or oil will make installation of piston pin easier.

Each piston is fitted with a plain fac-ed chrome top ring which may be in-stalled either side up. The plain faced cast iron, second compression ring may also be installed either side up. The third compression ring consists of four steel segments as shown in Fig. 69.

NOTE: Segments appear practically flat when not under compression; recommend-ed installation procedure is as follows: Grasp end and compress the segment as shown in the inset (Fig. 69) until ring ends slightly overlap. When compressed, ring ends will curl up, as shown, or down if segment is turned over. Ring ends should curl down on segment placed in bottom of groove, up on second segment, down on third segment and up on top segment. Space end gaps 180° apart for alternate segments after all are installed.

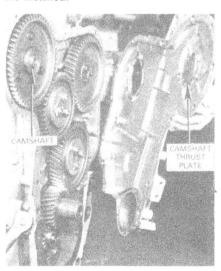

Fig. 67—Camshaft forward thrust is taken by plate attached to timing gear cover shown.

The fourth groove uses a six piece oil ring as shown in Fig. 70. Install the ex-pander first. Install the two lower rails, the spiral ring, then the two upper rails. Be sure to stagger gaps of the six parts of this ring. The lower groove contains a cast iron oil ring which may be install-ed either side up.

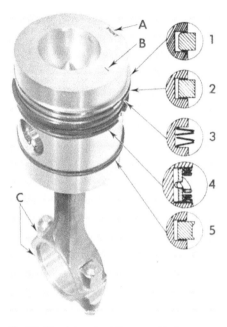

Fig. 68—The piston should be assembled correctly to rod as shown with rod correlation marks (C) on side indicated, cylinder mark (B) at location shown and "Front" or "F" mark at (A).

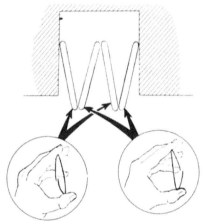

Fig. 69—The third ring is composed of four segments as shown. When ring segment is com-pressed by hand, ends of ring will point in direc-tion which inner portion slants.

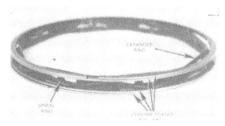

Fig. 70—View of the six-piece fourth ring show-ing correct assembly.

Piston ring specifications are as follows:

Ring End Gap

Top Ring . . . . .0.36-0.48 mm
(0.014-0.019 in.)
Second Ring . . . . .0.28-0.41 mm
(0.011-0.016 in.)
Third Ring
(Segments). . . . . .0.25-0.38 mm
(0.010-0.015 in.)
Fourth Ring
(Rails) . . . . .0.46-0.94 mm
(0.018-0.037 in.)
Bottom Ring . . . . .0.28-0.41 mm
(0.011-0.016 in.)

Side Clearance
in Groove
Top and Second
Rings . . . . .0.056-0.106 mm
(0.0022-0.0042 in.)
Third and Fourth
Rings . . . . .Not Applicable
Bottom Ring . . . . .0.051-0.102 mm
(0.0020-0.0040 in.)

Piston skirt diameter (measured at bottom of skirt and at right angle to piston pin bore) should be 91.326-91.351 mm (3.5955-3.5965 inches), and recommended diametral clearance in new cylinder bore is 0.13-0.18 mm (0.005-0.007 inch). Piston height relative to top surface of cylinder block must be checked when renewing pistons. Piston crown should be 0.076 mm (0.003 inch) below to 0.127 mm (0.005 inch) above surface of cylinder block. It may be necessary to machine top of new pistons to achieve this dimension.

Replacement (service) cylinder liners are a transitional fit, 0.025 mm (0.001 inch) loose to 0.025 mm (0.001 inch) tight, in cylinder block bores. Cylinder liners used in production are a slightly tighter press fit. Original cylinder liner inside diameter is 91.478-91.503 mm (3.6015-3.6025 inches). Renew liners if cylinder bore taper exceeds 0.15 mm (0.006 inch) or if out-of-round in excess of 0.10 mm (0.004 inch).

Use a suitable sleeve puller and installing tool when renewing cylinder liners. Make certain that cylinder bore is thoroughly cleaned and lightly lubricated before installing new liner. Press liner into cylinder bore until top of liner is flush to 0.10 mm (0.004 inch) below top surface of cylinder block. Allow installed liner to stabilize, then check for distortion which should not exceed 0.05 mm (0.002 inch). Inside diameter of liner is prefinished and should be 91.478-91.503 mm (3.6015-3.6025) inches after installation.

PISTON PINS

All Models So Equipped

62. The 31.744-31.750 mm (1.2498-1.2500 inches) diameter floating type piston pins are retained in piston bosses by snap rings. Pins are available in standard size only.

The renewable connecting rod bushing must be final sized after installation to provide a diametral clearance of 0.0127-0.0444 mm (0.00050-0.00175 inch) for the pin. Be sure the drilled oil hole in bushing is properly aligned with hole in top of connecting rod when installing new bushings.

The piston pin should have a 0.0025 mm (0.0001 inch) interference to 0.0006 mm (0.000025 inch) clearance in piston, which should be a hand press fit in piston after piston is heated to 40°-50° C (100°-120° F) in hot oil or water.

CONNECTING RODS AND BEARINGS

All Models So Equipped

63. Connecting rod bearings are precision type, renewable from below after removing oil pan and bearing caps. When renewing bearing shells, be sure that the projection engages milled slot in rod and cap and that the correlation marks are in register and face away from camshaft side of engine. Replacement rods should be marked with the cylinder number in which they are installed. Bearings are available in standard, as well as 0.25, 0.50 and 0.75 mm (0.010, 0.020 and 0.030 inch) undersizes.

Connecting rod bearings should have a diametral clearance of 0.064-0.101 mm (0.0025-0.0040 inch) on the 57.112-57.125 mm (2.2485-2.2490) inches

diameter crankpin. Recommended connecting rod side clearance is 0.25-0.38 mm (0.010-0.015 inch.)

Tighten the connecting rod nuts evenly to a torque of 54-61 N·m (40-45 ft.-lbs.)

CRANKSHAFT AND BEARINGS

All Models So Equipped

64. The crankshaft is supported in five precision type main bearings. To remove the rear main bearing cap, it is first necessary to remove the engine, clutch, flywheel, flywheel adapter and rear oil seal. The remainder of the main bearing caps can be removed after removing the oil pan.

Upper and lower bearing inserts are interchangeable on all except the front main bearing. When renewing the front bearing, make sure the correct half is installed in engine block. Bearing inserts are available in standard size and 0.25, 0.50 and 0.75 mm (0.010, 0.020 and 0.030 inch) undersizes.

Block and cap bores are machined with caps in place and caps cannot be interchanged. Block and caps are stamped with identification number and location number as shown in Fig. 71. When installing caps, cap identification number must be on same side of block as the block number. Bearing caps are aligned to block by ring dowels as shown at (A—Fig. 72).

Tighten the main bearing cap screws to a torque of 150-155 N·m (110-115 ft.-lbs.).

Crankshaft end play is controlled by renewable thrust washers (B and C—Fig. 72) at front and rear of rear main bearing. The cap half of thrust washer is prevented from turning by the tab which fits in a machined notch in cap. Block half of washer can be rolled from

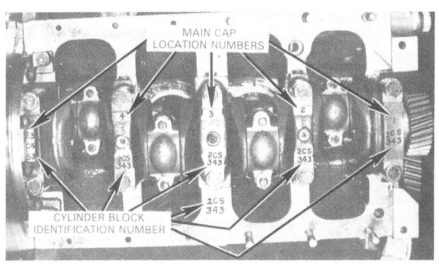

Fig. 71—Matched block and bearing caps bear a stamped identification number. Bearing caps are numbered front to rear as shown.

position when bearing cap is removed. Recommended crankshaft end play is 0.10-0.35 mm (0.004-0.014 inch). Thickness of standard size thrust washer is 3.125-3.175 mm (0.123-0.125 inch) when new. Thrust washers are also available in 0.18 mm (0.007 inch) oversize. Be sure grooved side of thrust washers face crankshaft.

Check the crankshaft journals against the following specifications:

Main Journal
Standard Diameter 69.812-69.825 mm
(2.7485-2.7490 in.)
Main Bearing
Clearance . . . . 0.08-0.13 mm
(0.003-0.005 in.)

Crankpin

Standard Diameter .57.112-57.125 mm
(2.2485-2.2590 in.)
Rod Bearing
Clearance . . . . 0.064-0.102 mm
(0.0025-0.0040 in.)

ENGINE ADAPTER PLATE

All Models So Equipped

65. The engine flywheel is housed in a cast iron adapter plate which is located to the engine block by two dowels and secured to block rear face by six cap screws. To obtain access to the rear main bearing, crankshaft or rear oil seal, it is first necessary to remove the adapter plate as follows:

Split tractor between engine and transmission as outlined in paragraph 148 and remove clutch and flywheel.

CAUTION: Flywheel is only lightly piloted to crankshaft flange and is not doweled. Use care when removing to prevent flywheel from falling and causing possible injury.

After flywheel is off, remove the cap screws securing adapter plate to engine block and tap plate free of locating dowels.

Install by reversing the removal procedure. Tighten flywheel retaining cap screws to a torque of 100-108 N·m (74-80 ft.-lbs.).

REAR OIL SEAL

All Models So Equipped

66. The asbestos rope type rear oil seal is contained in a two-piece seal retained at rear of engine block. The seal retainer can be removed after removing flywheel and adapter plate.

IMPORTANT: When rear oil leaks are present, check before and during disassembly to make sure crankshaft seal is actually leaking. Other points of possible leakage are the cup plug (P—Fig. 77) located in rear face of engine block to seal the main oil gallery, seal retainer gaskets, or lower retainer cap screws.

The rope type crankshaft seal is precision cut to length, and must be installed in retainer halves with 0.25-0.50 mm

Fig. 73—Crankshaft end play can be measured with feeler gage as shown.

(0.010-0.020 inch) of seal ends projecting from each end of retainer. To install the seal, clamp each half of retainer in a vise as shown in Fig. 74. Make sure seal groove is clean. Start each end of seal in groove with specified amount of free end protruding. Allow seal rope to buckle in the center until about 25 mm (1 inch) of each end is bedded in groove, work center of seal into position, then roll with a round bar as shown. Repeat the process with the other half of seal. Install seal retainer as outlined in paragraph 67.

67. Use a straightedge as shown in Fig. 76 when installing oil pan seal housing to make sure rear face of housing and block are flush. Coat both sides of retainer gasket and end joints of retainer halves with a suitable gasket cement. Coat shaft surface of rope seal with graphite grease. Install retainer halves and all cap screws loosely, then tighten clamp screws to 8 N·m (6 ft.-lbs.) torque before tightening the retaining cap screws gradually to 16 N·m (12 ft.-lbs.) torque. Then, finish tightening clamp bolts to 16 N·m (12 ft.-lbs.) torque. The four lower cap screws (C—Fig. 75) are

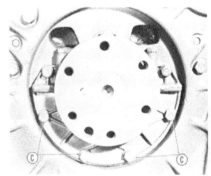

Fig. 75—Installed view of rear oil seal. The four cap screws (C) are sealed with aluminum washers.

Fig. 72—Renewable thrust washers (B and C) control crankshaft end play. Main bearing caps are positively located by ring dowels (A).

Fig. 74—Use a round bar to seat the rope type seal into retainer half. DO NOT trim ends of seal.

Fig. 76—When installing rear oil seal housing, use a straightedge as shown to make sure rear edge is flush with block rear face.

sealed with aluminum washers to prevent oil seepage at cap screw threads. Make sure aluminum washers are in good condition and properly installed.

FLYWHEEL

All Models So Equipped

68. To remove the flywheel, first separate the engine from transmission housing as outlined in paragraph 148 and remove the clutch. Flywheel is secured to crankshaft flange by six evenly spaced cap screws. To properly time flywheel to engine during installation, be sure that unused hole in flywheel aligns with untapped hole in crankshaft flange.

CAUTION: Flywheel is only lightly piloted to crankshaft. Use caution when unbolting flywheel to prevent flywheel from falling and causing possible injury.

The starter ring gear can be renewed after flywheel is removed. Heat new ring gear evenly to approximately 245° C (475° F), then install on flywheel with beveled end of teeth facing front of engine.

Check flywheel runout with a dial indicator after flywheel is installed. Maximum allowable flywheel runout must not exceed 0.025 mm (0.001 inch) for each 25 mm (1 inch) from flywheel centerline to point of measurement. Tighten the flywheel retaining cap screws to a torque of 100-108 N·m (74-80 ft.-lbs.) when installing the flywheel.

Fig. 77—To renew rear oil seal, the oil pan, flywheel and engine adapter plate must be removed as shown. Check the plug (P) in oil gallery for signs of leakage while adapter plate is off. Coat surface (S) on both left and right sides with sealer before assembling.

OIL PAN

All Models So Equipped

69. The heavy cast-iron oil pan serves as the tractor frame and attaching point for the tractor front support. To remove the oil pan, first drain cooling system and oil pan. Remove front axle and timing gear cover as outlined in paragraph 51. Unbolt power steering pump from engine. Remove oil pan attaching cap screws and stud nuts, then lower the oil pan away from engine block.

When installing oil pan, tighten the 5/16 inch fasteners to 26-28 N·m (19-21 ft.-lbs.) torque and the 7/16 inch fasteners to 69-77 N·m (51-57 ft.-lbs.) torque. Refer to paragraph 51 for installation of timing gear cover.

OIL PUMP

All Models So Equipped

70. The rotary type oil pump is mounted on front main bearing cap and driven from the crankshaft timing gear through an idler as shown in Fig. 78.

To remove the oil pump, first remove oil pan as outlined in paragraph 69. Pump can be removed as an assembly with front bearing cap or detached from cap as shown in Fig. 79 after removing idler gear.

Check clearance between inner rotor and annulus with a feeler gage as shown in Fig. 80. Clearance should be 0.013-

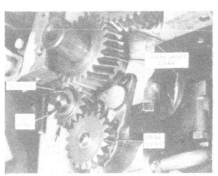

Fig. 78—Mounted view of engine oil pump and drive assembly with oil pan removed.

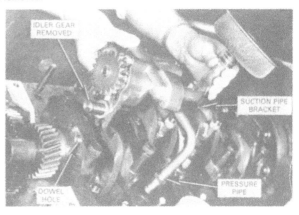

Fig. 79—Engine oil pump may be unbolted from front main bearing cap after removing idler gear.

0.063 mm (0.0005-0.0025 inch). Outer rotor (annulus) to body clearance (Fig. 81) should be 0.025-0.076 mm (0.001-0.003 inch). Check rotor end clearance with a straightedge and feeler gage as shown in Fig. 82. End clearance of inner rotor should be 0.038-0.076 mm (0.0015-0.0030 inch), and end clearance of outer rotor should be 0.013-0.063 mm (0.0005-0.0025 inch).

Except for the drive idler gear, relief valve plunger and relief valve spring, the oil pump is available only as an assembly. If rotor or body are scored or otherwise damaged, or fail to meet test specifications listed, renew the pump.

When reassembling, be sure that annulus is installed chamfered end first into pump body.

Fig. 80—Clearance between inner rotor and outer rotor (annulus) should be measured as shown with feeler gage.

Fig. 81—Clearance between outer rotor (annulus) and body should be measured with feeler gage.

Fig. 82—End clearance of inner rotor and outer rotor (annulus) should be within recommended limits.

71. RELIEF VALVE. The plunger type oil pressure relief valve (Fig. 83) is located in the oil pump body. The relief valve is set to open at 345-445 kPa (50-65 psi). All components of relief valve are renewable.

Relief valve spring should have a free length of 38.1 mm (1-1/2 inches). Spring should test 23.4-32.2 N (5-1/4 to 7-1/4 pounds) at 31.75 mm (1-1/4 inches).

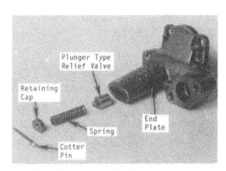

Fig. 83—Oil pump relief valve is located in the pump end plate.

PERKINS A4.236 and A4.248 DIESEL ENGINES AND COMPONENTS

Late production MF255 diesel models and all MF265 and MF270 diesel models are equipped with a Perkins A4.236 diesel engine. The four cylinder, direct injection diesel engine has a bore of 98.43 mm (3.875 inches), a stroke of 127 mm (5.0 inches) and a displacement of 3.87 L (236 cu. in.).

Models MF275 and MF290 are equipped with a Perkins A4.248 diesel engine. The four cylinder, direct injection diesel engine has a bore of 100.96 mm (3.975 inches), a stroke of 127 mm (5.0 inches) and a displacement of 4.06 L (248 cu. in.).

R&R ENGINE ASSEMBLY

All Models So Equipped

75. To remove engine assembly, first drain cooling system. Remove front grille panels, disconnect battery cables and remove hood. Disconnect headlight wires, radiator hoses, air cleaner hose and hydraulic cooler lines. Detach battery cable from starter motor, remove left rear side panel and disconnect power steering oil lines. Disconnect hoses from hydraulic oil cooler if so equipped. Disconnect radiator support bar. Support tractor with a jack at front of transmission housing. Place wedges between front axle and axle support to prevent tipping. Remove front end weights if so equipped. Support front axle and radiator so the front end can be moved forward when detached, then unbolt and separate front assembly from engine (Fig. 85).

Shut fuel off at tank, disconnect fuel gage sending unit wire, main fuel line and fuel return line, then remove tank and support plate. Disconnect main wiring harness from starter motor and water temperature gage sending unit.

Disconnect oil pressure gage line, tachometer drive cable, fuel shut-off cable and ether starting aid line (if so

equipped). Disconnect fuel line between filter and injection pump at injection pump end. Remove fuel return line between fuel filter and leak-off manifold.

Remove main fuel line between tank and fuel lift pump. Remove the two power steering oil lines (Fig. 87) between pump and control valve. Remove

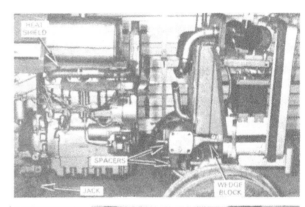

Fig. 85—View of tractor with front assembly rolled forward. Wedge blocks on each side between axle and front support prevent tipping sideways.

Fig. 86—View of engine left side. Be sure that tubes and wires are not caught when removing or installing engine.

Fig. 87—View of engine right side. Lines, cables and wires must be carefully positioned out of the way when removing and installing engine to prevent damage.

throttle rod. Disconnect exhaust pipe from models with low exhaust. On all models, be sure that all wires, linkage and tubes are positioned out of the way for engine removal. Attach hoist to lifting eyes of engine then remove screws attaching engine to transmission housing. Guide studs should be used while separating and rejoining engine to transmission.

Reassemble in reverse of removal procedure. Coat input shaft splines lightly with lithium base grease before assembling. Be sure that spacers (Fig. 85) are correctly positioned when reinstalling front end. Tighten engine to transmission attaching bolts to 75 N·m (55 ft.-lbs.) torque. Tighten axle support retaining bolts to 235 N·m (175 ft.-lbs.) torque.

CYLINDER HEAD

All Models So Equipped

76. REMOVE AND REINSTALL. To remove the cylinder head, drain cooling system and remove hood. Disconnect fuel gage sending unit wire, main fuel line and fuel return line, then remove fuel tank, heat shield and support plate. Disconnect upper radiator hose, air cleaner hose, water temperature gage sending unit wire and breather tube. Remove intake and exhaust manifold, fuel leak-off line, fuel leak-off manifold, high pressure fuel lines, injectors, rocker arm cover and rocker arm shaft assembly. Lift out all of the push rods, then loosen cylinder head stud nuts and retaining cap screws in reverse of order shown in Fig. 88.

Check flatness of cylinder head gasket surface using a straightedge and feeler gage. If surface is warped more than 0.10 mm (0.004 inch), cylinder head may be resurfaced. Up to 0.30 mm (0.012 inch) of material may be removed from cylinder head surface providing that injector nozzles do not protrude more than 4.44 mm (0.175 inch). Check for correct valve head recession as outlined in paragraph 77, and reseat valves as necessary after cylinder head is resurfaced.

Thoroughly clean mating surface of cylinder block and head, then position head gasket with "TOP FRONT FIT

DRY" marking toward top front. Install the head and tighten fasteners in sequence shown in Fig. 88 in 27 N·m (20 ft.-lbs.) increments until final torque of 135 N·m (100 ft.-lbs.) is obtained. Loosen rocker arm adjusting screws several turns to ensure valves do not contact the pistons, then install rocker shaft assembly and tighten retaining nuts to 33 N·m (24 ft.-lbs.) torque. Valve clearance cold should be 0.30 mm (0.012 inch). Refer to paragraph 81 for adjustment procedure.

Remainder of installation is reverse of removal procedure. Retorque cylinder head and adjust valve clearance after engine has reached operating temperature. Valve clearance hot should be 0.25 mm (0.010 inch) for all valves.

VALVES AND SEATS

All Models So Equipped

77. Intake and exhaust valves seat directly in cylinder head. Seat inserts are available for service installation. Valve face angle is 45° for intake and exhaust valves. Production valve seat angle is 46° for intake and exhaust; however, valve seats may be refaced to 45° angle for service. Intake valve seat has a 30° "flare" machined at top of seat. Desired seat width is 1.6-2.4 mm (1/16 to 3/32 inch) for intake and 2.4-2.8 mm (3/32 to 7/64 inch) for exhaust. Renew valve if bent, damaged or worn to point that valve head margin is less than 0.8 mm (1/32 inch).

Valve heads must be recessed a specified amount below surface of cylinder head. The recession (clearance) between top of valve head and cylinder head gasket surface can be measured using a suitable depth gage or a straightedge and feeler gage as shown in Fig. 89. Desired valve recession is 0.89-1.14 mm (0.035-0.045 inch) for intake valves and 1.07-1.32 mm (0.042-0.052 inch) for exhaust valves. If intake valve is recessed more than 1.55 mm (0.061 inch) or if exhaust valve is recessed more than 1.40 mm (0.055 inch), new valve and/or seat insert should be installed.

The cylinder head must be machined to provide appropriate size counterbore

for installation of valve seat inserts. Depth (D—Fig. 90) of intake counterbore should be 7.19-7.31 mm (0.283-0.288 inch), width (W) should be 51.22-51.24 mm (2.0165-2.0175 inches) and radius (R) should be 0.38 mm (0.015 inch) maximum. Exhaust counterbore depth (D) should be 9.525-9.652 mm (0.375-0.380 inch), width (W) should be 42.62-42.64 mm (1.678-1.679 inches) and radius (R) should be 0.38 mm (0.015 inch) maximum.

Chill insert in dry ice, then press insert squarely into cylinder head until it bottoms in counterbore. Reface valve seat as necessary to obtain valve recession within specified limits.

VALVE GUIDES

All Models So Equipped

78. Intake and exhaust valve guides are not renewable. Valves with 0.076, 0.38 and 0.76 mm (0.003, 0.015 and 0.030 inch) oversize stems are available for service if stem-to-guide clearance exceeds specifications. The valve guides must be reamed to provide desired clearance when fitting valves with oversize stems. Be sure to reface valve seats after reaming valve guides. Valve stem seals are used on both intake and exhaust valves.

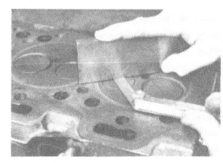

Fig. 89—Valve heads must be recessed the amount specified in text. Measure as shown using straightedge and feeler gage.

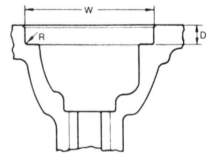

Fig. 90—Cylinder head must be machined to provide appropriate size counterbore for installation of valve seat inserts. Refer to text for specified dimensions.
D. Depth
R. Radius
W. Width

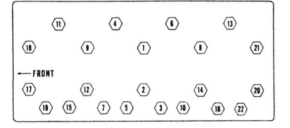

Fig. 88—Tighten the cylinder head stud nuts and retaining screws in the order shown. Loosen in reverse order to prevent warpage while removing.

Refer to the following specifications:

Guide ID-Standard

Intake and
  Exhaust . . . . .9.507-9.530 mm
            (0.3743-0.3752 in.)

Valve Stem OD-Standard

Intake . . . . .9.462-9.487 mm
            (0.3725-0.3735 in.)
Exhaust . . . . .9.449-9.468 mm
            (0.3720-0.3728 in.)

Valve Stem-to-Guide
  Clearance

Intake-Desired . . . . . .0.019-0.070 mm
            (0.00075-0.00275 in.)
  Wear Limit . . . . .0.13 mm
            (0.005 in.)
Exhaust-Desired . . . . .0.038-0.081 mm
            (0.0015-0.0032 in.)
  Wear Limit . . . . .0.15 mm
            (0.006 in.)

VALVE SPRINGS

All Models So Equipped

79. Valve springs, retainers and locks are interchangeable for intake and exhaust valves. Be sure to install valve springs with closely wound end (damper coils) toward cylinder head.

Renew valve springs that are rusted, discolored or distorted.

ROCKER ARMS

All Models So Equipped

80. The rocker arms and shaft assembly can be removed after removing the hood, fuel tank and rocker arm cover. Rocker arms should be kept in order as they are removed so they can be reinstalled in their original locations.

Desired diametral clearance between rocker arms and shaft is 0.025-0.089 mm (0.0010-0.0035 inch). Shaft diameter should be 19.012-19.037 mm (0.7485-0.7495 inch), and rocker arm bushing inside diameter should be 19.063-19.101 mm (0.7505-0.7520 inch). Bushings are renewable. Be sure to align oil hole in bushing with hole in rocker arm. New bushings must be reamed to final size after installation.

Rocker arms are right-hand and left-hand units and must be installed on shaft as shown in Fig. 91. Tighten rocker shaft support bracket retaining nuts to 33 N·m (24 ft.-lbs.) torque.

VALVE CLEARANCE

All Models So Equipped

81. The recommended valve clearance (tappet gap) with engine cold is 0.30 mm (0.012 inch) for both the intake and exhaust valves. Static setting of all valves can be made from just two crankshaft positions as follows:

Remove plug from left side of cylinder block and turn crankshaft until "TDC" timing mark (Fig. 92) on flywheel is aligned with timing pointer. Check the rocker arms for front and rear cylinders. If rear rocker arms are tight and front rocker arms have clearance, No. 1 piston is on compression stroke; adjust the four valves shown in Fig. 93. If front rocker arms are tight and rear rocker arms have clearance, No. 4 piston is on compression stroke; adjust the four valves shown in Fig. 94. After adjusting the indicated valves, turn crankshaft one complete turn until "TDC" timing mark (Fig. 92) is again aligned and adjust the remaining valves.

Recommended valve clearance with engine at normal operating temperature is 0.25 mm (0.010 inch) for both the intake and exhaust valves. Clearance may be adjusted following the procedure recommended for initial adjustment.

TIMING GEAR COVER

All Models So Equipped

82. To remove the timing gear cover, drain cooling system, remove front grille panels, disconnect battery cables and remove hood. Disconnect headlight wires, radiator hoses, air cleaner hose and hydraulic cooler lines. Detach battery cable from starter and power steering oil lines at upper left side of engine. Support tractor with a jack at front of transmission and place wedges between front axle and axle support to prevent tipping. Remove front weights if so equipped. Support front axle so that it can be moved forward when detached, then unbolt and separate front assembly from engine.

Remove fan belt, fan blades, crankshaft pulley and tachometer drive unit, then unbolt and remove timing gear cover.

Crankshaft front oil seal can be renewed at this time. Install seal in cover with sealing lip to rear, with front edge of seal recessed approximately 9.65-9.91 mm (0.380-0.390 inch) into seal bore when measured from front cover.

Timing gear cover is not doweled; use Special Tool MFN-747B or the crankshaft pulley as a pilot to properly align the oil seal when reinstalling the cover. Aluminum sealing washers must be installed on the four lower timing gear cover cap screws. Tighten timing gear cover retaining screws to 16-20 N·m (12-15 ft.-lbs.) and the crankshaft pulley retaining cap screw to a torque of 380-405 N·m (280-300 ft.- lbs.). Install the tachometer drive unit, then complete the assembly by reversing the disassembly procedure.

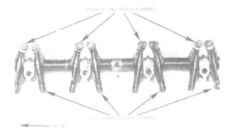

Fig. 91—View of rocker arms showing correct assembly of left and right hand units.

Fig. 92—View of flywheel timing mark aligned with pointer indicating number one cylinder is at "TDC". Number 4 cylinder is also at "TDC" at this time.

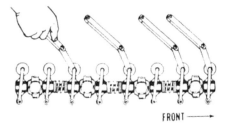

FRONT ——→

Fig. 93—With "TDC" flywheel timing marks aligned and No. 1 piston on compression stroke, adjust the indicated valves.

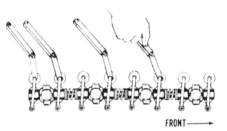

FRONT ——→

Fig. 94—With "TDC" flywheel timing marks aligned and No. 4 piston on compression stroke, adjust the indicated valves.

TIMING GEARS

All Models So Equipped

83. Fig. 95 shows a view of timing gear train with cover removed. Before attempting to remove any of the timing gears, first remove fuel tank, rocker arm cover and rocker arms to avoid the possibility of damage to pistons or valve train if camshaft or crankshaft should either one be turned independently of the other.

Timing gear backlash should be 0.08-0.15 mm (0.003-0.006 inch) between idler gear and any of its mating gears. Backlash between balancer idler gear and either of its mating gears should be 0.15-0.23 mm (0.006-0.009 inch). Replacement gears are available in standard size only. If backlash is not within the specified limits, renew gears, idler shafts or other items concerned.

NOTE: Because of the odd number of teeth on idler gear, all timing marks will not align every time No. 1 piston is at TDC on compression stroke. Complete set of timing marks will align only once in 18 crankshaft revolutions.

To remove the timing gears or time the engine, refer to the appropriate following paragraphs:

84. IDLER GEAR AND HUB. The timing idler gear (Fig. 95) should have a diametral clearance of 0.059-0.119 mm (0.0023-0.0047 inch) and end play of 0.10-0.20 mm (0.004-0.008 inch) on idler gear hub. The hub is a light press fit in timing gear housing, and can be loosened with a soft hammer if renewal is necessary. Due to uneven spacing of hub studs, hub can only be installed in one position.

The two flanged bushings in idler gear are renewable. Bushings must be reamed after installation to an inside diameter of 50.795-50.818 mm (1.9998-2.0007 inches). Tighten idler gear retaining screws to a torque of 41 N·m (30 ft.-lbs.), then bend lock plate around heads of cap screws.

85. CAMSHAFT GEAR. The camshaft gear (Fig. 95) is pressed and keyed to shaft and retained by a special cap screw, tab washer and retaining plate. Use a suitable puller to remove the gear.

Use the retaining plate and cap screw to draw gear into position when reinstalling. Do not attempt to drive gear onto camshaft as the plug located in cylinder block at rear of camshaft will be damaged. Tighten retaining cap screw to a torque of 68 N·m (50 ft.-lbs.), then lock in place by bending tab washer. Time the gears as outlined in paragraph 88.

86. INJECTION PUMP DRIVE GEAR. The injection pump drive gear (Fig. 95) is retained to pump adapter by three cap screws. When installing the gear, align dowel pin with slot (Fig. 96) in adapter hub, then install the retaining cap screws. The injection pump drive gear and adapter are supported by the injection pump rotor bearings.

87. CRANKSHAFT GEAR. The crankshaft timing gear (Fig. 95) is keyed to the shaft and is a transition fit, 0.025 mm (0.001 inch) tight to 0.025 mm (0.001 inch) loose on shaft. It is usually possible to remove the gear using two small pry bars to move the gear forward. Remove timing gear housing and engine balancer, then use a suitable puller if gear cannot be removed with pry bars.

Be sure that oil slinger is installed between crankshaft gear and timing gear cover with cup side toward front. Be sure that engine balancer gears are correctly timed to crankshaft gear as outlined in paragraph 88.

88. TIMING THE GEARS. To install and time the gears, first install crankshaft, camshaft and injection pump gears as outlined in paragraphs 85, 86 and 87 with timing marks to front. Turn the shafts until timing marks (Fig. 95) point toward idler gear hub, then install idler gear with all marks aligned as shown. Secure idler gear as in paragraph 84.

TIMING GEAR HOUSING

All Models So Equipped

89. To remove the timing gear housing, first remove timing gears as outlined in paragraphs 84 through 87 and the injection pump as in paragraph 126. Remove power steering pump if not previously removed. If oil pan has not been removed, remove the four front oil pan cap screws. Remove the cap screws

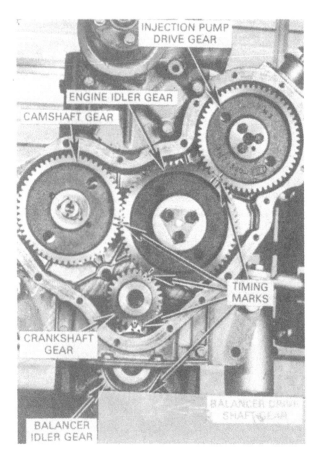

Fig. 95—View of timing gears installed and marks aligned for A4.236 and A4.248 diesel engines.

Fig. 96—Correct installation of injection pump drive gear is simplified by the dowel pin which fits in machined slot in pump drive shaft.

securing timing gear housing to block front face and lift off the housing as shown in Fig. 97.

Timing gear housing must be removed before camshaft can be withdrawn. Install by reversing the removal procedure. Oil pan must be installed and tightened before tightening the cap screws securing timing gear housing to block front face.

CAMSHAFT

All Models So Equipped

90. To remove the camshaft, first remove timing gear housing as outlined in paragraph 89. Secure cam followers (tappets) in their uppermost position, remove the fuel lift pump, then withdraw camshaft and front thrust washer.

The camshaft runs in three bearings. The front bearing bore contains a pre-sized renewable bushing, while the center and rear journals ride directly in machined bores in engine block (Fig. 98).

Camshaft end play of 0.10-0.40 mm (0.004-0.016 inch) is controlled by a thrust washer which is located behind timing gear housing in a recess machined in front face of cylinder block. Thrust washer should protrude above face of

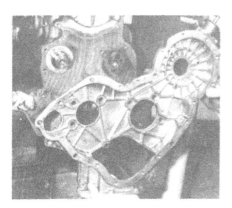

Fig. 97—Timing gear housing must be removed as shown before camshaft can be withdrawn.

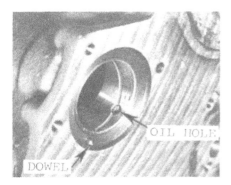

Fig. 98—Front face of engine block with camshaft removed, showing front camshaft bushing and thrust washer locating dowel. The other two camshaft bores are not bushed.

cylinder block 0.66-0.79 mm (0.026-0.031 inch). Thickness of thrust washer when new is 5.49-5.54 mm (0.216-0.218 inch). Thrust washer should be renewed if camshaft end play exceeds 0.51 mm (0.020 inch).

Camshaft specifications are as follows:

Front Journal
Journal OD . . . . . . .50.711-50.736 mm
           (1.9965-1.9975 in.)
Bearing Bore ID . . . . .50.80-50.83 mm
           (2.000-2.001 in.)
Clearance Limits . . . .0.064-0.114 mm
           (0.0025-0.0045 in.)

Center Journal
Journal OD . . . . . . .50.457-50.482 mm
           (1.9865-1.9875 in.)
Bearing Bore ID . . . . .50.55-50.60 mm
           (1.990-1.992 in.)
Clearance Limits . . . .0.064-0.140 mm
           (0.0025-0.0055 in.)

Rear Journal
Journal OD . . . . . . .49.949-49.974 mm
           (1.9665-1.9675 in.)
Bearing ID . . . . .50.04-50.09 mm
           (1.970-1.972 in.)
Clearance Limits . . . .0.064-0.140 mm
           (0.0025-0.0055 in.)

CAM FOLLOWERS

All Models So Equipped

91. The mushroom type cam followers (tappets) operate directly in machined bores in cylinder block. Cam followers can be removed through bottom of cylinder block after removing camshaft and oil pan. Identify cam followers as they are removed so they can be reinstalled in their original locations if reused.

The 18.99-19.01 mm (0.7475-0.7485 inch) diameter cam followers are furnished in standard size only and should have a diametral clearance of 0.038-0.095 mm (0.0015-0.0038 inch) in block bores. Cam followers should be renewed if scored, pitted or excessively worn. Always renew cam followers whenever camshaft is renewed.

ROD AND PISTON UNITS

All Models So Equipped

92. Connecting rod and piston units are removed from above after removing cylinder head, oil pan, engine balancer and rod bearing caps. Be sure to remove carbon deposit and ring ridge (if present) from top of cylinder before pushing piston out of cylinder.

Cylinder numbers are stamped on the connecting rod and cap. When reinstalling, make sure correlation numbers are

in register and face away from camshaft side of engine. Each piston is stamped with a number corresponding to its cylinder location. The piston crown is also stamped with the word "FRONT" or letter "F" to indicate correct installation (Fig. 99).

Note that two different types of connecting rod cap nuts, each with different tightening torque values, have been used. Early models used cadmium plated (bright finish) nuts while later models use phosphated (dull black finish) nuts. Threads should be lightly oiled and nuts tightened to the following torque values: 108 N·m (80 ft.-lbs.) for cadmium plated nuts and 129 N·m (95 ft.-lbs.) for phosphated nuts.

PISTONS, SLEEVES AND RINGS

All Models So Equipped

93. The aluminum alloy, cam ground pistons are supplied in standard size only and are available in a kit consisting of piston, pin and rings for one cylinder. The toroidal combustion chamber is offset in piston crown and piston is marked "FRONT" or "F" for proper assembly as shown at (A— Fig. 99). Refer

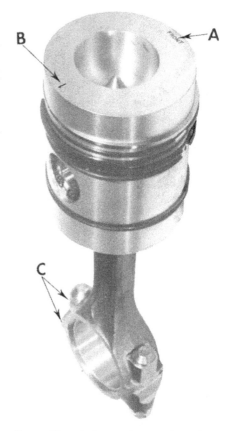

Fig. 99—View of piston and connecting rod correctly assembled with "F" or "FRONT" mark (A), cylinder number (B) and rod correlation numbers (C).

to Fig. 100 or 101 and the following for installation of piston rings.

Five piston rings are used on A4.236 models. Production piston rings (P) are used with new sleeves and service ring sets (S) should be used in worn sleeves.

On production rings sets (P), the top ring (1) is chrome plated, parallel faced, cast iron and may be installed with either side up. The second and third compression rings (2 and 3) are equipped with groove (step) internally which should be toward top with "BTM" marking down. The cast iron oil control rings (4 and 5) may be installed with either side up.

On service piston ring sets (S), the chrome plated top ring (1) should be installed with groove in outer diameter and side marked "TOP" toward top of piston. The cast iron second ring (2) may be installed with either side up. The third ring (3) should be installed with groove (step) in inner diameter toward top of piston. The three-piece, fourth ring (4) should be installed by positioning expander first, then staggering the end gaps of the two rails. The fifth ring (5) may be installed with either side up.

Four piston rings are used on early Model A4.248 engines (prior to engine S.N. U839080J), while three piston rings are used on late production A4.248 engines (starting with engine S.N. U839080J).

On early A4.248 engines, the chrome plated top compression ring is barrel faced and may be installed either side up. The second and third compression rings are chrome plated with internal groove (step) which should be toward top with side marked "BTM" down. Install the spring into fourth groove making sure that latch pin enters both ends of spring, then install cast iron oil ring over spring with end gap positioned 180° from spring latch pin.

On late Model A4.248 engines with three rings, the molybdenum top compression ring (Fig. 101) is barrel faced

with top face internally stepped and externally chamfered. Second compression ring is cast iron with a tapered face. Install with internal stepped groove up and external step down. The spring loaded oil control ring may be installed either side up.

To check piston ring grooves for wear on all engines, use a feeler gage to measure ring side clearance with new ring installed in groove. Renew piston assembly if side clearance exceeds specified limits.

Check ring end gap using a feeler gage with rings positioned in unworn area of sleeve at top of cylinder. Refer to the following for piston ring specifications:

A4.236 Engine (5 Rings)
Ring Side Clearance—
  Compression Rings
  (1, 2 & 3) 0.05-0.10
    (0.002-0.004 in.)
  Oil Control Rings
  (4 & 5) . . . . 0.06-0.11 mm
    (0.0025-0.0045 in.)
Ring End Gap—

  Compression Ring (1) . 0.41-0.86 mm
    (0.016-0.034 in.)
  Compression Rings
  (2 & 3) . . . . 0.30-0.76 mm
    (0.012-0.030 in.)
  Oil Control Rings
  (4 & 5) . . . . 0.30-0.76 mm
    (0.012-0.030 in.)

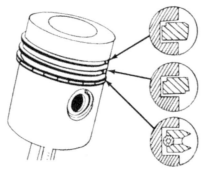

Fig. 101—Drawing of late production A4.248 three-ring piston showing correct installation of piston rings.

A4.248 Engine (4 Rings)
Ring Side Clearance—
  Compression Rings
  (1, 2 & 3) . . . . 0.05-0.10 mm
    (0.002-0.004 in.)
  Oil Control
  Ring (4) . . . . 0.06-0.11 mm
    (0.0025-0.0045 in.)
Ring End Gap—
  Compression Ring (1) . 0.41-0.86 mm
    (0.016-0.034 in.)
  Compression Ring
  (2 & 3) . . . . 0.30-0.76 mm
    (0.012-0.030 in.)
  Oil Control
  Ring (4) . . . . 0.30-0.91 mm
    (0.012-0.036 in.)

A4.248 Engine (3 Rings)
Ring Side Clearance—
  Compression Rings
  (1 & 2) . . . . 0.04-0.07 mm
    (0.0017-0.0027 in.)
  Oil Control
  Ring (3) . . . . 0.05-0.10 mm
    (0.0020-0.0038 in.)

On all models, check installed piston height at TDC when any new piston is installed (Fig. 102). Top of piston must be 0.41-0.62 mm (0.016-0.024 inch) above top face of cylinder block on A4.236 engines and 0.08-0.25 mm (0.003-0.010 inch) above top of cylinder block on A4.248 engines. If piston height exceeds specified limit it will be necessary to machine piston crown to obtain specified installed height.

Cylinder sleeves used in original production are 0.03-0.08 mm (0.001-0.003 inch) interference fit in cylinder block bores. Fit of service cylinder sleeves in cylinder block is from 0.03 mm (0.001 inch) tight to 0.03 mm (0.001 inch) loose. Original inside diameter of production cylinder sleeves is 98.48-98.50 mm (3.877-3.878 inches for A4.236 engines and 101.05-101.07 mm (3.9785-3.9795 inches) for A4.248 engines. Sleeves should be renewed if cylinder bore taper exceeds 0.2 mm (0.008 inch). A suitable sleeve removal and installing tool is required to renew sleeves.

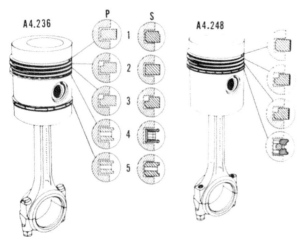

Fig. 102—Measure installed height of new pistons as shown. It may be necessary to machine piston crown to obtain specified piston height.

Fig. 100—Drawing of pistons with cross section of piston rings. Production rings for A4.236 engines are shown at (P) and service rings (used with worn sleeves) are shown at (S). Early production A4.248 four-ring piston is also shown. Refer to Fig. 101 for late production three-ring piston used on A4.248 engine.

Thoroughly clean and inspect cylinder block bores and sleeves before installing new sleeves, as even the slightest burr or dirt can cause distortion of new sleeve when pressed into the block. Lubricate cylinder bores with a thin lubricant prior to installing sleeves. Press sleeve into cylinder block until sleeve flange bottoms in cylinder block counterbore. When correctly installed, top of sleeve flange should be from 0.10 mm (0.004 inch) below to 0.10 mm (0.004 inch) above top surface of cylinder block. Note that early production sleeves had a fire ring that protruded above the sleeve flange. The late style sleeves (without fire ring) may be used to replace early style sleeves when renewing complete set of sleeves. Do not intermix early and late style sleeves.

Allow time for sleeve to stabilize, then check to make sure that sleeve bore is not distorted. Service cylinder sleeves are prefinished and should not require additional machining after installation. Inside diameter of sleeves after installation should be 98.50-98.52 mm (3.878-3.879 inches) for A4.236 engines and 101.09-101.12 mm (3.980-3.981 inches) for A4.248 engines.

PISTON PINS

All Models So Equipped

94. The floating type piston pins are retained in piston bosses by snap rings. The 34.920-34.925 mm (1.3748-1.3750 inches) diameter pins are available in standard size only. The pins are a transitional fit, 0.0 mm (0.0 inch) interference to 0.01 mm (0.0004 inch) clearance, in piston bosses. Heating piston in hot water or oil will make removal and installation of pins easier.

The renewable connecting rod bushing must be final sized after installation to provide a diametral clearance of 0.019-0.043 mm (0.0008-0.0017 inch) for the pin. Be sure that oil hole in bushing is aligned with hole in top of connecting rod when installing new bushings.

Piston pin retaining rings should be renewed whenever they are removed.

CONNECTING RODS AND BEARINGS

All Models So Equipped

95. Connecting rod bearings are precision insert type, renewable from below after removing oil pan, engine balancer and bearing caps. When renewing bearing shells, be sure that the projection engages milled slot in rod and cap and that the correlation marks on rod and cap are in register and face away from camshaft side of engine. Replacement rods should be marked with the cylinder number in which they are installed. Bearings are available in standard as well as 0.25, 0.50 and 0.75 mm (0.010, 0.020 and 0.030 inch) undersizes.

Connecting rod bearings should have a diametral clearance of 0.031-0.079 mm (0.0012-0.0031 inch) on the 63.470-63.490 mm (2.4988-2.4996 inch) standard diameter crankpin. Recommended connecting rod side clearance is 0.22-0.37 mm (0.0085-0.0145 inch).

Note that two different types of connecting rod cap nuts, each with different tightening torque values, have been used. Early models used cadmium plated (bright finish) nuts while later models use phosphated (dull black finish) nuts. Threads of connecting rod bolts should be lightly oiled and the nuts tightened to the following torque values: 108 N·m (80 ft.-lbs.) for cadmium plated nuts and 129 N·m (95 ft.-lbs.) for phosphated nuts.

CRANKSHAFT AND BEARINGS

All Models So Equipped

96. The crankshaft is supported in five precision type main bearings. To remove the rear main bearing cap, it is first necessary to remove the engine, clutch, flywheel and real oil seal. The remainder of the main bearing caps can be removed after removing the oil pan and engine balancer.

Upper and lower main bearing inserts are not interchangeable, the upper (block) half being slotted to provide pressure lubrication to crankshaft and connecting rods. Inserts are interchangeable in pairs for all journals except the center main bearing. Bearing inserts are available in standard size and 0.25, 0.50 and 0.75 mm (0.010, 0.020 and 0.030 inch) undersizes.

Crankshaft end play is controlled by renewable thrust washers at front and rear of center main bearing. The cap half of thrust washer is prevented from turning by the tab which fits in machined notch in cap. Block half of washer can be rolled from position when bearing cap is removed. Recommended crankshaft end play is 0.10-0.38 mm (0.004-0.015 inch). Thrust washers are available in 0.19 mm (0.0075 inch) oversize as well as standard thickness. One set (top and bottom half) of oversize thrust washers may be installed on one side (front or rear) in combination with standard size set at other side to provide oversize adjustment of 0.19 mm (0.0075 inch). Two sets of oversize thrust washers may be used to provide adjustment of 0.38 mm (0.015 inch).

Check crankshaft journals and bearings against the values listed below:

Main Journal Diameter
Standard . . . . .76.16-76.18 mm
(2.9985-2.9990 in.)
Maximum Allowable
Journal Wear . . . . .0.04 mm
(0.0015 in.)
Center Main Journal
Width . . . . .44.145-44.221 mm
(1.738-1.741 in.)
Crankpin Diameter .63.47-63.49 mm
(2.4990-2.4995 in.)
Diametral Clearance
Main Bearings . . . . . . .0.05-0.11 mm
(0.002-0.004 in.)
Crankpin Bearings . . .0.03-0.08 mm
(0.0015-0.0030 in.)

Main bearing caps are numbered 1 through 5 with number 1 at front of engine. Each cap is also marked with a serial number which matches number stamped on bottom of cylinder block. Tighten main bearing cap bolts to 244 N·m (180 ft.- lbs.) torque. Lightly coat sealing surfaces of cylinder block bridge piece with nonhardening sealer. Make sure bridge piece is flush with rear surface of cylinder block as shown in Fig. 104. Refer to appropriate paragraph 97 or 98 for rear oil seal installation.

CRANKSHAFT REAR OIL SEAL

Engines with the letter ''L'' in suffix of engine serial number use a one-piece, lip type crankshaft rear oil seal. Early production engines without the letter ''L'' in engine serial number use a two-piece rope type rear oil seal.

All Models With One-Piece Lip Type Seal

97. The lip type rear oil seal is contained in a one-piece seal retainer attached to rear face of engine block. The seal retainer and seal can be removed after splitting tractor between engine and transmission as outlined in paragraph 148 and removing engine clutch and flywheel.

Remove old seal from retainer, then thoroughly clean retainer. Press new seal squarely into retainer making sure that seal lip will be toward the engine when installed. Note that latest style seal also has a dust deflector lip that protrudes rearward.

Inspect sealing surface of crankshaft flange for wear or damage. If flange is grooved, new seal may be pressed slightly further into retainer so it will contact an unworn area of flange. If a satisfactory sealing area on crankshaft cannot be obtained, it may be possible to machine worn sealing area of flange, but do not alter area on which flywheel locates.

Lubricate lip of seal and the crankshaft flange with clean oil. Install seal and retainer assembly and tighten retaining cap screws evenly to 24 N·m (18 ft.-lbs.) torque.

All Models With Two-Piece Rope Type Seal

98. The rope type rear oil seal is contained in a two-piece retainer attached to rear face of cylinder block. The seal retainer and seal can be removed after splitting tractor between engine and transmission as outlined in paragraph 148 and removing engine clutch and flywheel.

The rope type crankshaft seal is precision cut to length, and must be installed in retainer halves with 0.25-0.50 mm (0.010-0.020 inch) of seal end protruding from each end of retainer groove. Do not trim the seal. To install the seal, clamp each retainer half in a vise as shown in Fig. 105. Make sure seal groove is clean. Start each end in groove with the specified amount of seal protruding. Allow seal rope to buckle in the center until about 25 mm (1 inch) of each end is imbedded in groove, work center of seal into position, then roll with a round bar as shown. Repeat the process with other half of seal retainer.

When installing cylinder block bridge piece, insert end seals as shown in Fig. 103. Use a straightedge as shown in Fig. 104 to make sure bridge piece is flush with rear face of cylinder block.

Coat both sides of retainer gasket and end joints of retainer halves with a suitable gasket cement. Coat surface of rope seal with graphite grease. Install retainer halves and cap screws loosely.

Tighten the two clamp bolts (Fig. 106) first to 8 N·m (6 ft.-lbs.) torque, then tighten seal retainer cap screws evenly to 16 N·m (12 ft.-lbs.), and finally tighten clamp bolts to 16 N·m (12 ft.-lbs.) torque.

ENGINE BALANCER

Early Models (Prior to Engine S.N. U880190K)

99. OPERATION. The Lanchester type engine balancer consists of two unbalanced shafts which rotate in opposite directions at twice crankshaft speed. The inertia of the shaft weights is timed to cancel out natural engine vibration, thus producing a smoother running engine. The balancer is correctly timed when the balance weights are at their lowest point when pistons are at TDC and BDC of their stroke.

The balancer unit is driven by the crankshaft timing gear through an idler gear attached to balancer frame. The engine oil pump is mounted at rear of balancer frame and driven by the balancer shaft. Refer to Figs. 107 through 109.

Fig. 104—Use a straightedge to align bridge piece flush with rear surface of cylinder block.

100. REMOVE AND REINSTALL. The balancer assembly can be removed after removing the oil pan and mounting cap screws. Engine oil is pressure fed through balancer frame and cylinder block. Balancer frame bearings are also pressure fed.

When installing balancer with engine in tractor, timing marks will be difficult to observe without removing gear cover. The balancer assembly can be safely installed as follows:

Turn crankshaft until No. 1 and No. 4 pistons are at the exact bottom of their stroke. Remove balancer idler gear (5—Fig. 109) if necessary, and reinstall with single punch-marked tooth of idler gear meshed between the two marked teeth on weight drive shaft as shown in Fig. 107. Install balancer frame with balance weights hanging normally (flat surfaces of weights aligned and facing upward). If carefully installed, timing will be correct although timing marks at front of engine may not be aligned. (Marks align only occasionally when engine is running.) If engine is mounted in stand or tractor front end is removed, timing marks can be observed by removing timing gear cover.

NOTE: Balancer can be safely installed with No. 1 and 4 pistons at either TDC or BDC. BDC is selected because interference between connecting rod and balance weights can give warning if unit is badly out of time. Also, alignment of timing marks is not essential but is a convenience for original engine assembly.

With balancer correctly installed, tighten the retaining cap screws to a torque of 54 N·m (40 ft.-lbs.) and complete the assembly by reversing the removal procedure.

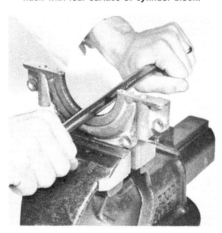

Fig. 103—Cylinder block bridge is equipped with a seal at each end. A nonhardening sealing compound should also be applied to sealing surfaces of the bridge.

Fig. 105—Use a round bar to seat the asbestos rope seal in each retainer half. Refer to text for details.

Fig. 106—Rear view of cylinder block showing two-piece oil seal retainer. Refer to text for mounting bolt tightening sequence.

When reassembling balancer, use Loctite 242 on threads of screws retaining gears (14—Fig. 109) to balance weights and on threads of set screws (S) retaining balance weights to shafts. Make certain that flat surfaces of weights are aligned and facing upward as shown in Fig. 108, then install idler gear aligning single timing mark on idler gear with double timing marks on balancer drive gear. Tighten idler gear retaining nut to 54 N·m (40 ft.-lbs.) torque.

ENGINE BALANCER

Late Models (With Engine S.N. U880190K and After)

101. OVERHAUL. Refer to Fig. 109 for an exploded view of balancer frame and associated parts. To disassemble the removed balancer unit, unbolt and remove oil pump housing (20) and associated parts; and idler gear (5) and associated parts. Set screws (S) retaining balance weights (15 and 16) are installed using "Loctite 242 Threadlocker." Loosen screws, then push balance shafts (11 and 12) forward out of frame and weights.

NOTE: Use care when removing shafts not to allow keys (13) to damage frame bushings.

Check the balancer components against the following specifications:

102. GENERAL. The late style engine balancer consists of two shaft and weight assemblies (2 and 3—Fig. 110) which rotate in opposite directions at twice crankshaft speed. The balancer weights will be positioned at their lowest point (flat surfaces of weights upward) each time the pistons are at TDC and BDC of their stroke if properly timed.

The balancer unit is driven by the crankshaft gear through an idler gear attached to balancer frame. The engine oil pump (16) is mounted in the center of balancer frame and is driven by the balancer drive shaft.

Fig. 108—Assembled view of removed early style engine balancer and oil pump unit. Refer to Fig. 109 for exploded view.

Drive Shaft (11)
Journal Diameter—
    Front ....31.725-31.737 mm
                 (1.2490-1.2495 in.)
    Rear ....25.375-25.387 mm
                 (0.9990-0.9995 in.)
Bearing Clearance—
    Front ....0.051-0.107 mm
                 (0.0020-0.0042 in.)
    Rear ....0.046-0.089 mm
                 (0.0018-0.0035 in.)

Driven Shaft (12)
Journal Diameter—
    Front..........31.710-31.725 mm
                 (1.2485-1.2490 in.)
    Rear ....25.367-25.380 mm
                 (0.9987-0.9992 in.)
Bearing Clearance—
    Front ....0.064-0.122 mm
                 (0.0025-0.0048 in.)
    Rear ....0.054-0.096 mm
                 (0.0021-0.0038 in.)

Idler Gear (5)
Bushing ID ....38.100-38.140 mm
                 (1.5000-1.5015 in.)
Diametral Clearance
    on Hub..........0.025-0.081 mm
                 (0.0010-0.0032 in.)
End Play..........0.20-0.35 mm
                 (0.008-0.014 in.)
Backlash..........0.15-0.23 mm
                 (0.006-0.009 in.)

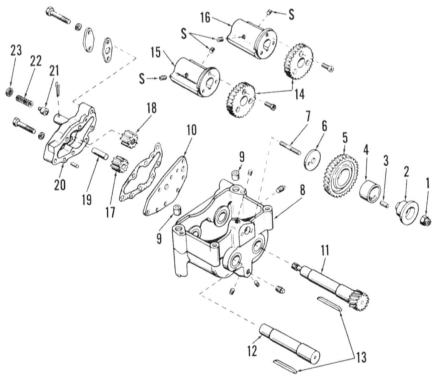

Fig. 109—Exploded view of engine balancer and engine oil pump used on early production engines (prior to engine S.N. U880190K).

1. Locknut	7. Stud	13. Key	19. Shaft
2. Hub	8. Frame	14. Gear	20. Pump body
3. Dowel	9. Ring dowels	15. Balance weight	21. Valve piston
4. Bushing	10. Plate	16. Balance weight	22. Valve spring
5. Idler gear	11. Drive shaft	17. Pump gear	23. Cap
6. Washer	12. Driven shaft	18. Pump gear	S. Set screw

103. REMOVE AND REINSTALL. The balancer assembly can be removed after removing the oil pan and balancer mounting cap screws.

To reinstall and "time" balancer to crankshaft, first rotate crankshaft until No. 1 and 4 pistons are at TDC of their stroke. With balancer weights hanging downward (flat surfaces of weights facing upward), position balancer unit on bottom of cylinder block meshing idler gear with crankshaft gear. Tighten balancer mounting cap screws to 54 N·m (40 ft.-lbs.) torque.

104. OVERHAUL. Refer to Fig. 110 for an exploded view of balancer assembly. To disassemble the removed balancer unit, remove cap screw retaining idler gear (9) and hub (11) to balancer frame. Hold the drive shaft gear (12), then remove nut (19) from end of shaft. Withdraw drive shaft and balancer drive gear (18). Remove cover plate (15). Unbolt and remove oil pump (16) from the frame. Remove cap screws retaining oil transfer plate (4) and end cover (6), then separate plate and cover from frame. Remove the balance weights (2 and 3). Press bushings (1) from end cover and frame if necessary.

Remove plugs (P) noting their position for reassembly, and flush out lubricating oil passages. Examine all parts for wear or damage and renew if necessary. Refer to the following specifications:

Drive Shaft (12)
Journal OD—
Front . . . . .28.562-28.575 mm
(1.1245-1.1250 in.)
Rear . . . . .23.787-23.800 mm
(0.9365-0.9370 in.)
End Play . . . . .0.17-0.35 mm
(0.0067-0.0137 in.)

Balance Weights (2 & 3)
Shaft OD . . . . .38.054-38.069 mm
(1.4982-1.4988 in.)
Bushing ID . . . . . . .38.133-38.174 mm
(1.501-1.503 in.)
Shaft to Bushing
Diametral Clearance0.064-0.0120 mm
(0.0025-0.0047 in.)
Backlash . . . . .0.10-0.20 mm
(0.004-0.008 in.
End Play . . . . .0.186-0.377 mm
(0.0073-0.0148 in.)

Idler Gear (9)
Hub OD . . . . .38.09-38.10 mm
(1.4995-1.5000 in.)
Thrust Washer
Thickness . . . . .4.14-4.29 mm
(0.163-0.169 in.)
End Play . . . . .0.08-0.23 mm
(0.003-0.009 in.)

When renewing balance weight bushings (1—Fig. 110), press new bushings into balancer housing until rear end of bushings is 2.50-2.72 mm (0.098-0.107 inch) below machined surface of housing bore. Press bushings into end cover (6) until front end of bushings is 2.75-3.25 mm (0.108-0.125 inch) below machined surface of cover.

If drive shaft bearings are being renewed, press rear bearing (17) into housing bore until rear end of bearing is 2.0-3.0 mm (0.079-0.118 inch) below rear face of housing. Press front bearing (13) into housing bore until it is 2.5-3.0 mm (0.098-0.118 inch) below front face of housing.

Install oil pump in housing and tighten mounting cap screws to 33 N·m (24 ft.-lbs.) torque.

To assemble and correctly time balancer weights, proceed as follows: Position weights in balancer housing with flat surfaces of weights aligned and facing upward. The edge of the flat on right-hand weight (3) should align with scribed line on inner wall of balancer housing. Insert drive shaft (12) through the oil pump and position shaft so smaller diameter hole (A—Fig. 111) in

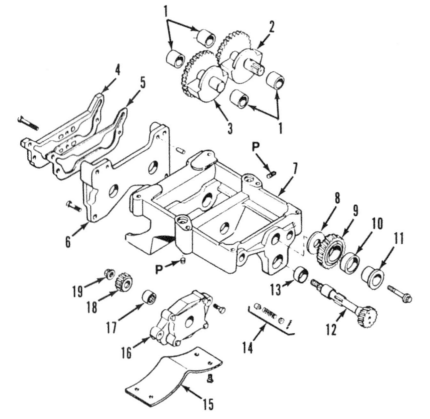

Fig. 110—Exploded view of engine balancer and engine oil pump used on late production engines (engine S.N.U880190K and after).

P. Plugs
1. Bushings
2. Balancer weight
3. Balancer weight
4. Oil transfer plate
5. Gasket
6. End cover
7. Balancer frame
8. Thrust washer
9. Idler gear
10. Bearing
11. Hub
12. Drive shaft
13. Bearing
14. Oil pressure relief valve
15. Cover plate
16. Oil pump assy.
17. Bearing
18. Balancer drive gear
19. Locknut

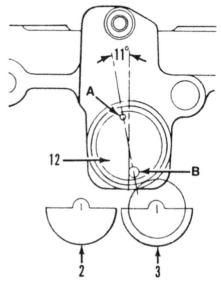

Fig. 111—To time balance weights, position flat surfaces of weights (2 and 3) upward and in line with each other. Install drive shaft (12) with small diameter hole (A) in end of shaft positioned 11° left of vertical as shown.

end of drive shaft/gear (12) is located approximately 11° left of vertical as shown. While holding balance weights and drive shaft in timed position, install balancer weight drive gear (18—Fig. 110) onto drive shaft. Apply Loctite 242 to threads of locknut (19), then tighten nut to 95 N·m (70 ft.-lbs.) torque.

Install the end cover (6) and tighten mounting cap screws to 54 N·m (40 ft.-lbs.) torque. Install oil transfer plate (4) with a new gasket and tighten mounting cap screws to 33 N·m (24 ft.-lbs.) torque. Note that plugs (P) for lubricating oil passages are of different lengths and must be installed in correct locations. The plugs have different shaped socket heads (hexagon socket on short plug, square socket on long plug), and correct assembly location of plugs is indicated by the raised hexagon symbol (H—Fig. 112) and square symbol (S) on rear of oil transfer plate (4).

Install idler gear (9—Fig. 110) and hub (11) assembly. Tighten retaining cap screw to 75 N·m (55 ft.-lbs.) torque.

OIL PAN

All Models So Equipped

105. The heavy cast iron oil pan serves as part of tractor frame and as attaching point for tractor front support. To remove the oil pan, first drain the oil. Support tractor under transmission housing, remove cap screws securing oil pan to cylinder block, front support and transmission housing and lower the oil pan from cylinder block.

Install by reversing the removal procedure. Tighten retaining screws to 19 N·m (14 ft.-lbs.) torque. Tighten oil pan to transmission housing cap screws to 75 N·m (55 ft.- lbs.) torque and oil pan to front axle support to 270 N·m (200 ft.-lbs.) torque.

OIL PUMP

Early Models (With Oil Pump Mounted on Rear of Balancer Housing)

106. The gear type oil pump is mounted on engine balancer frame and driven by balancer shaft as shown in Fig. 109. Oil pump can be removed after removing oil pan.

Examine pump body (20—Fig. 109), gears (17 and 18) and plate (10) for wear or scoring and renew any parts that are questionable. Pump gears must be renewed as a pair. Check parts against the following values:

Thickness of oil pump gears should be 25.35-25.40 mm (0.998-1.000 inch). Depth of pump body cavity should be 25.375-25.480 mm (0.999-1.002 inch).

End clearance of gears in pump body with a 0.10 mm (0.004 inch) gasket installed should be 0.08-0.20 mm (0.003-0.008 inch). Radial clearance of gears in body bores should be 0.05-0.23 mm (0.002-0.009 inch).

Late Models (With Oil Pump Mounted in Center of Balancer Housing)

107. The gerotor type oil pump is mounted in the center of engine balancer housing and driven by the balancer drive shaft. To remove oil pump (16—Fig. 110), first remove oil pan and engine balancer as outlined in paragraph 103. Remove balancer idler gear (9). Remove nut (19) and balancer drive gear (18) from drive shaft (12), then withdraw drive shaft from oil pump and balancer housing. Remove oil pump mounting cap screws and lift out the oil pump assembly.

Remove end cover from oil pump body and remove inner and outer rotors. Check rotors, pump body and end cover for wear or other damage. Individual parts are not available for pump repair other than relief valve assembly (14). Check parts against the following values to determine if pump assembly should be renewed.

Inner Rotor to Outer
  Rotor Clearance . . .0.025-0.150 mm
                   (0.001-0.006 in.)
Outer Rotor to
  Body Clearance . . . . . .0.31-0.44 mm
                   (0.012-0.017 in.)
Inner and Outer Rotor
  End Clearance . . . .0.025-0.076 mm
                   (0.001-0.003 in.)

When installing pump, tighten mounting cap screws to 33 N·m (24 ft.-lbs.) torque. Install drive shaft (12) making certain that small hole (A—Fig. 111) in end of shaft/gear is positioned as shown and flat surfaces of balancer weights (2 and 3) are facing upward. Install balancer drive gear on end of drive shaft. Apply Loctite 242 to threads of

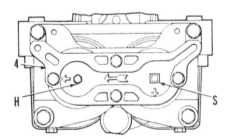

Fig. 112—The raised hexagon symbol (H) and square symbol (S) on rear of lubricating oil transfer plate (4) indicate correct installation position of plugs in lubricating oil passages. Refer to text.

locknut (19—Fig. 110), then tighten nut to a torque of 95 N·m (70 ft.-lbs.). Reinstall idler gear (9) assembly and tighten retaining cap screw to 75 N·m (55 ft.-lbs.) torque.

Reinstall balancer and oil pump assembly as outlined in paragraph 103.

OIL PRESSURE RELIEF VALVE

All Models So Equipped

108. The plunger type relief valve is located in oil pump body as shown in appropriate Fig. 109 or 110. The valve is set to open at approximately 415 kPa (60 psi). Normal operating oil pressure is 210-380 kPa (30-55 psi) with engine at normal operating temperature and running at full speed.

FLYWHEEL

All Models So Equipped

109. To remove the flywheel, first separate engine from transmission housing as outlined in paragraph 148 and remove the clutch. Flywheel is secured to crankshaft flange by six cap screws.

The flywheel ring gear may be removed at this time. To remove ring gear, a 5 mm (3/16 inch) hole may be drilled partially through ring gear at the root of any tooth. Then, use a chisel above the drilled hole to split the ring gear.

CAUTION: Cover chisel point with a cloth to protect against flying fragments as ring gear may disintegrate when struck.

An alternate method of removal is to heat ring gear with a torch to expand it, then drive gear off the flywheel.

Heat new ring gear to a temperature of 245° C (475° F) using an oven. Heating new gear with a torch is not recommended as excessive heat can ruin the heat treatment of gear. Install ring gear on flywheel with chamfered side of teeth facing front of flywheel. Allow ring gear to air cool.

To properly locate flywheel on crankshaft, align the seventh (unused) hole in flywheel with untapped hole in crankshaft flange. Tighten retaining cap screws evenly to 108 N·m (80 ft.-lbs.) torque.

Use a dial indicator to check run out of clutch face on flywheel. Push crankshaft fully forward to remove end play when making this check. Flywheel face should run true within 0.025 mm (0.001 inch) per 25 mm (1 inch) of flywheel radius, measured from center of crankshaft to contact point of dial indicator stylus. If run out is excessive, remove flywheel and check for burrs, foreign material or flywheel warpage.

GASOLINE FUEL SYSTEM

CARBURETOR

All Gasoline Models

110. ADJUSTMENT. Zenith Model 13337 carburetor is used on MF255 gasoline tractors (Perkins AG4.212 engine); Zenith 267LJ10 carburetor is used on MF265 gasoline tractors (Perkins AG4.236 engine). Initial adjustment is 1 to 1-1/4 turns open for idle mixture adjustment needle (Fig. 115). Final adjustment should be accomplished after engine has reached operating temperature. Adjust idle speed to 725-755 rpm after idle mixture is set. Location of the stop screw is shown in Fig. 116.

111. OVERHAUL. The two Zenith carburetors used are serviced similarly, but differences will be noted where necessary. To disassemble the removed carburetor, first clean outside with a suitable solvent. Remove the screws retaining throttle body (4—Fig. 117) to fuel bowl (13) and remove fuel bowl. Remove float shaft (8), float and inlet valve needle.

NOTE: Float shaft (8) is a tight fit in slotted side of hinge bracket and should be removed from opposite side.

Remove venturi (10), inlet valve seat and idle jet (6) from throttle body. Remove idle adjusting needle (5). Remove fuel shut-off solenoid (12), discharge jet (15), main jet (11) and well vent jet (14) from fuel bowl. Remove throttle and choke valves, shafts and packing.

Discard all gaskets and packing and clean remainder of parts in a suitable carburetor cleaner. Rinse in clean mineral solvent and blow out passages in body and bowl with compressed air. Renew all gaskets and packing and any other parts which are worn or damaged.

Refer to Fig. 117 and following specifications for the two different carburetors.

Zenith 13337

Venturi (10)		21 mm
Main Jet (11A) . . .	No. 27	1.35 mm
Idle Jet (6)	No. 16	0.80 mm
Idle Air		
Restriction	No. 31	1.55 mm
Main Discharge		
Jet (15)	No. 55	2.75 mm
Well Vent (14)	No. 16	0.80 mm
Fuel Valve		
Seat (7)	No. 35	1.75 mm

Zenith 267LJ10

Venturi (10)		23 mm
Main Jet (11B) . . .	No. 31	1.55 mm
Idle Jet (6)	No. 17	0.85 mm
Idle Air		
Restriction	No. 38	1.9 mm
Main Discharge		
Jet (15)	No. 70	3.5 mm
Well Vent (14)	No. 22	1.1 mm
Fuel Valve		
Seat (7)	No. 35	1.75 mm

Assemble by reversing the disassembly procedure, using new gaskets and packing. Install throttle valve so beveled edges will fit throttle body bore with throttle closed, with side of throttle plate farthest from mounting flange aligned with idle port. Adjust float height to 4.75-5.13 mm (0.187-0.202 inch) measured from gasket to nearest edge of float as shown in Fig. 118. Drill bits in 3/16 inch and 13/64 inch size can be used as Go and No-Go gages as shown. If adjustment is required, carefully bend float arms using needlenose pliers or a bending tool, keeping the two halves of float parallel and equal. Check idle mixture and idle speed adjustments after installation as outlined in paragraph 110.

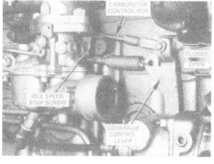

Fig. 115—View of carburetor installed on Perkins AG4.212 engine (MF255 tractor). The fuel shut-off solenoid is installed on angle as shown. Idle speed stop screw is at location indicated.

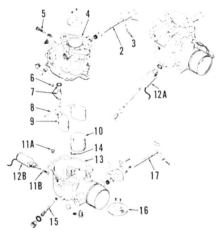

Fig. 117—Exploded view of Zenith carburetor. Main jet is located at (12A) and fuel shut-off valve at (11A) for 13337 carburetor and at (11B and 12B) for 267LJ10 carburetor.

2. Throttle shaft
3. Idle speed screw
4. Throttle body
5. Idle mixture needle
6. Idle jet
7. Inlet needle valve
8. Float shaft
9. Float
10. Venturi
11. Main jet
12. Shut-off solenoid
13. Float chamber
14. Well vent jet
15. Discharge jet
16. Choke valve

Fig. 116—View of carburetor installed on Perkins AG4.236 engine (MF265 tractor). The fuel shut-off valve is installed in horizontal position as shown. Idle mixture needle is at position indicated.

Fig. 118—Float height may be set using 3/16 inch and 13/64 inch drill bits as GO and NO-GO gages.

GOVERNOR

All Gasoline Models

112. ADJUSTMENT. Recommended governed engine speeds are as follows:

Low idle . . . . 725-775 rpm
High idle (no-load) . . . 2225-2275 rpm
Loaded speed . . . . 2000 rpm

To adjust the governor, first adjust the length of rod (1—Fig. 119) to 498.5 mm (19-5/8 inches) measured from end of spring pin to centerline of rod end. Adjust length of rod (2) to 140.5 mm (5-17/32 inches) measured between centerline of ball joints. Start engine and adjust length of rod (1) to provide high idle (no-load) speed of 2225-2275 rpm. If throttle lever creeps or will not maintain a set position, adjust the pressure on friction disc by tightening nut (3).

113. OVERHAUL. Governor arm shaft, bearings and associated parts can be inspected and/or overhauled by unbolting and removing governor housing (19—Fig. 120). To remove shaft and weight unit, it is first necessary to remove timing gear cover as outlined in paragraph 18 and governor gear as in paragraph 22. Distributor drive unit which carries governor weights can then be removed as outlined in paragraph 28. Shaft and weight unit is only available as an assembly.

Refer to Fig. 120 for a disassembled view of governor unit. The governor shaft (11) should have diametral clearance of 0.023-0.061 mm (0.0008-0.0024 inch) in bore of housing (7) and clearance of 0.102-0.140 mm (0.0040-0.0055 inch) in bushing (16). Recommended governor shaft end play is 0.10-0.20 mm (0.004-0.008 inch). Distributor drive shaft (10) should have 0.025-0.076 mm (0.001-0.003 inch) clearance in bushing bore. Recommended distributor drive shaft end play is 0.25-1.06 mm (0.010-0.042 inch). Control shaft (22) should have 0.025-0.076 mm (0.001-0.003 inch) diametral clearance in housing (19).

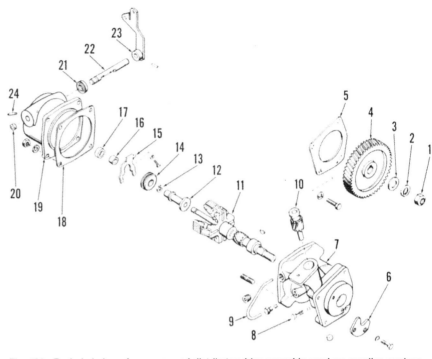

Fig. 120—Exploded view of governor and distributor drive assembly used on gasoline engines.

1. Nut
2. Lockwasher
3. Plain washer
4. Drive gear
5. Gasket
6. Thrust plate
7. Drive body
8. Distributor drive gear retaining pin
9. Oil line
10. Distributor drive gear
11. Drive shaft assy.
12. Sleeve
13. Snap ring
14. Thrust bearing
15. Yoke
16. Bushing
17. Seal
18. Gasket
19. Housing
20. Plug
21. Seal
22. Shaft
23. Lever
24. Shaft retaining pin

DIESEL FUEL SYSTEM

The diesel fuel system consists of three basic units: the fuel tank and filters, injection pump and injector nozzles. When servicing any unit associated with diesel fuel system, the maintenance of absolute cleanliness is of utmost importance. Of equal importance is the avoidance of nicks or burrs on any of the working parts.

FUEL FILTER AND BLEEDING

All Models So Equipped

114. OPERATION AND MAINTENANCE. Refer to Fig. 121 for a schematic view of fuel flow through filter, lift pump and injection pump. A much greater volume of fuel is cir-

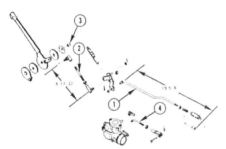

Fig. 119—Drawing of gasoline engine throttle linkage. Lengths of rods (1 and 2) should be adjusted to specified dimensions. Nut (3) adjusts lever friction. Refer to text.

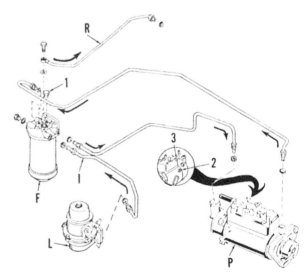

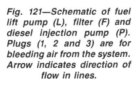

Fig. 121—Schematic of fuel lift pump (L), filter (F) and diesel injection pump (P). Plugs (1, 2 and 3) are for bleeding air from the system. Arrow indicates direction of flow in lines.

culated within the system than is burned in the engine, the excess serving as a coolant and lubricant for the injection pump.

Fuel enters the filter through inlet line (I—Fig. 121) from the fuel lift pump (L). Filtered fuel is fed to the injection pump (P) as shown by arrows. The transfer pump incorporated in the injection pump has the primary purpose of charging the delivery plungers which distribute the high-pressure fuel to the injectors. A greater quantity of fuel is pumped than is actually required to run the engine. The excess serves to cool and lubricate the pump, then enters the return line to filter (F) where it is recirculated. Line (R) returns excess fuel to the tank.

Inspect glass bowl at bottom of filter daily and drain off any dirt or water accumulation. Renew filter at 250 hour intervals or sooner if loss of engine power is evident. If evidence of substantial water or other contamination exists, source of contamination should be found and corrected before resuming operation of tractor.

115. BLEEDING. Air must be bled from fuel system whenever fuel tank has run dry, fuel filter has been renewed or any fuel lines have been disconnected. To bleed system, make sure that tank shut-off valve is open and all fuel line connections are tight. Loosen air vent plug (1—Fig. 121) on top of fuel filter housing. Operate priming lever on fuel lift pump (L) until air free fuel flows from vent plug, then tighten plug.

NOTE: If fuel lift pump does not pump fuel and no resistance can be felt when actuating priming lever, rotate engine crankshaft to reposition fuel pump actuating lobe on camshaft.

Move fuel shut-off control to RUN position. Loosen vent screws (2 and 3) on injection pump. Actuate fuel lift pump until fuel flows from bottom bleed screw (2), then tighten screw. Continue to operate fuel lift pump until fuel flows from top bleed screw (3), then tighten screw.

NOTE: Air in governor housing relieved by bleed screw (3) will not prevent tractor from starting and running properly; however, condensation in the trapped air can cause rusting of governor components and eventual pump malfunction. Do not fail to bleed governor housing even though the tractor starts and runs properly.

Operate manual lever approximately ten extra strokes after tightening vent plug (3) to expel any air remaining in bleed back lines.

With fuel supply system bled, push in the stop button, partially open throttle lever and attempt to start the tractor. If tractor fails to fire, loosen compression nut at all injector nozzles and turn engine over with starter until fuel escapes from all loosened connections. Tighten compression nuts and start engine.

FUEL LIFT PUMP

All Models So Equipped

116. The fuel lift pump shown in Fig. 123 is type used on MF255 diesel models. All other diesel models are equipped with fuel lift pump shown in Fig. 122. All of the pumps are driven by an eccentric on the engine camshaft. Output delivery pressure should be 35-55 kPa (5-8 psi) for Model MF255; 42-69 kPa (6-10 psi) for all other models.

Prior to disassembling fuel lift pump, scribe a mark across fuel pump body (8) and valve housing (4) to ensure correct alignment when reassembling. Turn diaphragm (7) 90° to disengage it from actuator lever (9). Inspect all parts and renew as necessary. Note that inlet and outlet valves (6) are not available

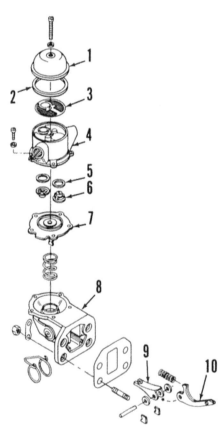

Fig. 122—Exploded view of fuel lift pump used on all diesel models except Model MF255.

1. Cover	6. Check valve
2. Gasket	7. Diaphragm
3. Strainer screen	8. Pump body
4. Valve housing	9. Actuator link
5. Gasket	10. Rocker arm

separately on pump shown in Fig. 122. A repair kit containing the diaphragm and other related components is available for all pumps.

INJECTOR NOZZLES

All diesel models are equipped with C.A.V. multi-hole injector nozzles which extend through the cylinder head to inject fuel charge directly into the combustion chamber.

All Models So Equipped

117. TESTING AND LOCATING A FAULTY NOZZLE. If rough or uneven engine operation, or misfiring indicates a faulty injector, the defective unit can usually be located as follows:

With engine running at the speed where malfunction is most noticeable (usually low idle speed), loosen the compression nut on high pressure line for each injector in turn and listen for a change in engine performance. The faulty injector is the one that least affects the running of the engine when its line is disconnected.

If a faulty nozzle is found and considerable time has elapsed since the injectors have been serviced, it is recommended that all nozzles be removed and serviced or that new or reconditioned units be installed. Refer to the following paragraphs for removal and test procedure.

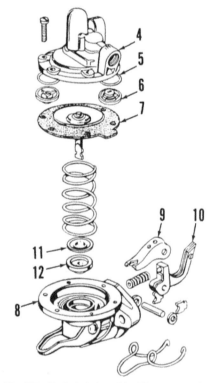

Fig. 123—Exploded view of fuel lift pump used on MF255 diesel models. Refer to Fig. 122 for legend except for seal (11) and retainer (12).

118. REMOVE AND REINSTALL. Before loosening any fuel lines, thoroughly clean the lines, injectors and surrounding area. Disconnect fuel leak-off line and pressure line and immediately cap all openings to prevent entry of dirt into fuel system. Remove the two stud nuts and withdraw injector unit from cylinder head.

NOTE: If injector is difficult to remove, apply penetrating oil and use a suitable puller (such as MFN330) to pull injector from cylinder head.

Thoroughly clean the nozzle recess in cylinder head before reinstalling injector unit. It is important that seating surface be free of even the smallest particle of carbon or dirt which could cause the injector unit to be cocked and result in blowby. No hard or sharp tools should be used in cleaning. Do not reuse copper sealing washer located between injector nozzle and cylinder head, always install a new washer. Each injector should slide freely into place in cylinder head without binding. Make sure that dust seal is reinstalled and tighten the retaining stud nuts evenly to a torque of 16 N·m (12 ft.-lbs.). After engine is started, examine injectors for blowby, making the necessary corrections before releasing tractor for service.

119. TESTING. A complete job of testing and adjusting the injector requires the use of special test equipment.

WARNING: Fuel leaves the injector nozzle with sufficient force to penetrate the skin. Keep exposed portions of your body clear of nozzle spray when testing.

Only clean, approved testing oil should be used in the tester tank. The nozzle should be tested for opening pressure, seat leakage, back leakage and spray pattern. When tested, the nozzle should open with a sharp popping or buzzing sound, and cut off quickly at end of injection with a minimum of seat leakage and controlled amount of back leakage.

Before conducting the test, operate tester lever until fuel flows, then attach the injector. Close the valve to tester gage and pump tester lever a few quick strokes to purge air from nozzle and to be sure nozzle valve is not plugged.

NOTE: Spray pattern is not symmetrical with centerline of nozzle tip. The apparently irregular location of nozzle holes (See Figs. 124 and 125) is designed to provide the correct spray pattern in the combustion chamber.

120. OPENING PRESSURE. Open the valve to tester gage and operate tester lever slowly while observing gage reading. Opening pressure should be 17240 kPa (2500 psi). If opening pressure is not as specified, remove the injector cap nut (1—Fig. 126) and turn adjusting sleeve (5) as required to obtain the recommended pressure.

NOTE: When adjusting a new injector or an overhauled injector with new pressure spring (6), set the pressure at 17720 kPa (2570 psi) to allow for initial pressure loss.

121. SEAT LEAKAGE. The nozzle valve should not leak at pressure less than 15860 kPa (2300 psi). To check for leakage, actuate tester lever slowly and as the gage needle approaches 15860 kPa (2300 psi), observe the nozzle tip. Hold the pressure at 15860 kPa (2300 psi) for ten seconds; if drops appear or if nozzle tip is wet, the valve is not

Fig. 124—Nozzle holes (arrows) are not located an equal distance from nozzle tip.

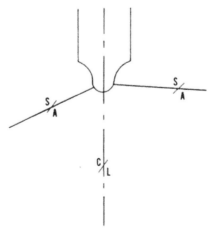

Fig. 125—Nozzle spray pattern is not symmetrical with centerline of nozzle tip.

seating and injector must be disassembled and overhauled as outlined in paragraph 124.

122. BACK LEAKAGE. If nozzle seat as tested in paragraph 121 was satisfactory, check the injector and connections for wetness which would indicate leakage. If no visible external leaks are noted, bring gage pressure to 15515 kPa (2250 psi), release the lever and observe the time for gage pressure to drop from 15515 kPa (2250 psi) to 10345 kPa (1500 psi). For a nozzle in good condition, this time should not be less than six seconds. A faster pressure drop would indicate a worn or scored nozzle valve piston or body, and the nozzle assembly should be renewed.

NOTE: Leakage of the tester check valve or connections will cause a false reading, showing up in this test as excessively fast leakback. If all injectors tested fail to pass the test, the tester rather than the unit should be suspected as faulty.

123. SPRAY PATTERN. If leakage and pressure are as specified when tested as outlined in paragraphs 120 through 122, operate the tester handle several times while observing spray pattern. Four finely atomized, equally spaced, conical sprays should emerge from nozzle tip, with equal penetration into the surrounding atmosphere.

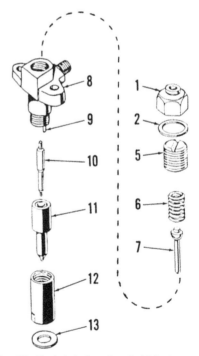

Fig. 126—Exploded view of typical injector nozzle and holder assembly.

1. Cap nut
2. Gasket
5. Adjusting sleeve
6. Spring
7. Valve spindle
8. Nozzle holder
9. Dowel pin
10. Nozzle valve
11. Nozzle body
12. Nozzle nut
13. Seat washer

If pattern is uneven, ragged or not finely atomized, overhaul the nozzle as outlined in paragraph 124.

NOTE: Spray pattern is not symmetrical with centerline of nozzle tip; refer to Fig. 125.

124. OVERHAUL. Hard or sharp tools, emery cloth, grinding compound, or other than approved solvents or lapping compounds must never be used. An approved nozzle cleaning kit is available through any C.A.V. Service Agency and other sources.

Wipe all dirt and loose carbon from exterior of nozzle and holder assembly. Refer to Fig. 126 and proceed as follows:

Secure the nozzle in a soft-jawed vise or holding fixture and remove cap nut (1), then back off the adjusting sleeve (5) to completely unload the pressure spring (6). Remove the nozzle cap nut (12) and nozzle body (11).

Nozzle valve (10) and body (11) are matched assemblies and must never be intermixed. Place all parts in a clean calibrating oil or diesel fuel as they are removed. Clean the exterior surfaces with a brass wire brush. Soak in an approved carbon solvent, if necessary, to loosen hard carbon deposits. Rinse the parts in clean diesel fuel or calibrating oil immediately after cleaning to neutralize the carbon solvent and prevent etching the polished surfaces.

Clean the pressure chamber of nozzle tip using the special reamer as shown in Fig. 127. Clean the spray holes in nozzle with a wire probe held in a pin vise as shown in Fig. 128. Use a 0.24 mm

(0.009 inch) diameter cleaning wire for MF255, MF265 and MF270 models and a 0.26 mm (0.010 inch) diameter cleaning wire for MF275 and MF290 models. The cleaning wire should protrude from pin vise only far enough to pass through spray holes, approximately 2 mm (1/16 inch), to prevent bending and breaking. Rotate pin vise without applying undue pressure until each hole is clear. Clean valve seats by inserting small end of brass valve seat scraper into nozzle and rotating tool. Reverse the tool and clean upper chamfer using large end. Refer to Fig. 129. Use the hooked scraper to clean annular groove in top of nozzle body (Fig. 130). Use the same hooked tool to clean the internal fuel gallery.

With the above cleaning accomplished, back flush the nozzle by installing the reverse flusher adapter on injector tester and nozzle body in adapter, tip end first. Secure with the knurled adapter nut and insert and rotate the nozzle valve while flushing.

Reclean all parts by rinsing thoroughly in clean diesel fuel or calibrating oil. Make certain that nozzle valve (10—Fig. 126) slides freely into nozzle body (11). If valve sticks, reclean nozzle assembly and recheck for burrs or carbon deposits. If valve still does not slide freely, renew nozzle assembly. Be sure lapped pressure faces of nozzle body and holder (8) are clean and free of scratches or other damage (Fig. 131).

Reassemble injector while wet with diesel fuel. Tighten nozzle cap nut (12—Fig. 126) to a torque of 68 N·m (50 ft.- lbs.). Do not overtighten as distortion may cause nozzle valve to stick.

Retest and adjust the assembled injector as outlined in paragraphs 119 through 123.

INJECTION PUMP

The injection pump is a completely sealed unit. No service work of any kind should be attempted on the pump or governor unit without the use of special pump testing equipment and special training. Inexperienced or unequipped service personnel should never attempt to overhaul a diesel injection pump.

All Diesel Models

125. ADJUSTMENT. The slow idle stop screw (Fig. 132) should be adjusted with engine warm and running to provide the recommended slow idle speed of 725-775 rpm. Also check to make sure that governor arm contacts the slow idle screw and high speed screw when throttle lever is moved to slow and fast positions. Also check to make sure that stop lever arm moves fully to operating position when stop button is pushed in, and shuts off the fuel to injectors when stop button is pulled.

The high speed stop screw is set at the factory and adjustment is sealed. Governed speed under full load should be 2000 rpm on all tractors except Model MF290 which should be 2200 rpm. High idle (no-load) speed should not exceed 2185 rpm on all models except MF290 which should not exceed 2380 rpm.

Refer to paragraph 127 for injection pump timing adjustment.

126. REMOVE AND REINSTALL. Prior to removing injection pump, thoroughly wash the pump and connections with clean diesel fuel or solvent. Do not wash injection pump with cold water while engine is running as seizure of close fitting parts may result.

Disconnect throttle control rod and engine shut-off control cable from injection pump. Shut off fuel supply at tank. Disconnect fuel inlet, outlet and high

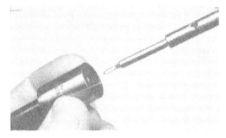

Fig. 127—Clean the pressure chamber in nozzle tip using the special reamer as shown.

Fig. 128—Clean the spray holes in nozzle tip using a pin vise and wire probe slightly smaller than orifice.

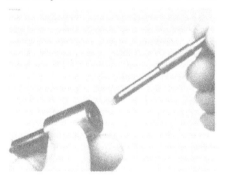

Fig. 129—Clean the valve seats using brass scraper as shown.

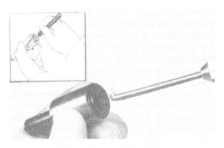

Fig. 130—Use the hooked scraper to clean the annular groove as shown in inset.

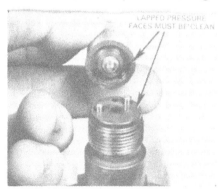

Fig. 131—Be sure lapped surfaces are clean and that locating dowels are perfectly aligned when nozzle body is reinstalled.

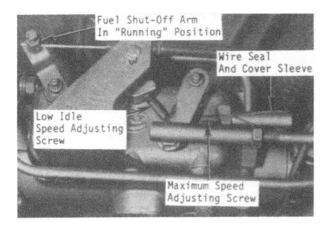

Fig. 132—View of typical injection pump linkage showing location of low speed and high speed adjustment screws.

should be aligned with timing mark at center of timing hole.

To check pump timing on all other models, proceed as follows: Shut off fuel and remove pump timing window (W—Fig. 134) from side of injection pump. Remove flywheel timing plug from left side of transmission adapter plate. Turn crankshaft in normal direction of rotation until the letter "C" timing marks on injection pump rotor is aligned with the straight edge of internal snap ring as shown in Fig. 138. At this point, the 23° BTDC mark on flywheel should be aligned with timing window indicator as shown in Fig. 137 for MF255, MF265 and

pressure lines from pump and immediately plug all openings to prevent entry of dirt. Remove the inspection cover from front of timing gear cover, then remove the three cap screws retaining injection pump drive gear to pump shaft. Remove stud nuts attaching pump to timing gear housing and withdraw pump assembly.

The injection pump drive shaft contains a milled slot (S—Fig. 133) which engages a dowel pin in injection pump drive gear. The drive gear cannot become unmeshed from timing idler gear without removal of timing gear cover. Therefore, injection timing is not disturbed by removal and installation of injection pump.

Normal installation of injection pump can be accomplished without reference to crankshaft timing marks or internal marks on injection pump. Engage slot (S—Fig. 133) in pump drive shaft with dowel pin in pump drive gear, and align timing marks (T—Fig. 134) scribed on pump flange and timing gear housing. Tighten pump stud nuts to a 27 N·m (20 ft.-lbs.) torque.

Complete installation of pump by reversing the removal procedure. Bleed air from system as outlined in paragraph 115. Check injection pump static timing, if necessary, as outlined in paragraph 127.

127. INJECTION PUMP STATIC TIMING. To check pump timing for Model MF255 equipped with AD4.203 engine, proceed as follows: Shut off the fuel supply. Remove timing window cover (W—Fig. 134) from side of injection pump. Remove flywheel timing plug from left side of transmission adapter housing. Turn crankshaft in normal direction of rotation until the letter "B" timing mark on injection pump rotor is aligned with the straight edge of internal snap ring as shown in Fig. 136. At this point, the 26° BTDC timing mark on flywheel (Fig. 135)

Fig. 134—Timing marks (T) must be aligned when injection pump is installed. Pump internal timing marks are located behind timing window cover (W).

Fig. 136—View of injection pump timing marks for Model MF255 equipped with AD4.203 engine showing "B" timing mark properly aligned with square end of internal snap ring. Refer to text.

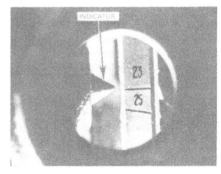

Fig. 137—The 23° mark on flywheel is used for timing injection for MF255, MF265 and MF270 models with A4.236 engine. Correct timing for MF275 and MF290 models with A4.248 engine is 24° which is between the two marks. Refer to text.

Fig. 133—Slot (S) in pump shaft must align with dowel in hub of injection pump drive gear when pump is installed.

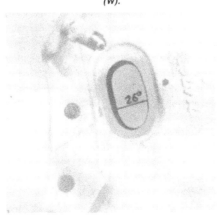

Fig. 135—The 26° BTDC mark on flywheel is used for timing injection for MF255 models equipped with AD4.203 engine.

Fig. 138—View of injection pump timing marks for all models equipped with A4.236 or A4.248 engine showing "C" timing mark properly aligned with square end of snap ring. Refer to text.

MF270 tractors equipped with A4.236 engine. On MF275 and MF290 tractors, the indicator pointer should be halfway between the 23° and 25° marks which will be 24° BTDC.

On all models, the mounting holes in pump mounting flange are elongated to permit minor timing adjustments. If pump housing is moved, be sure to scribe appropriate timing marks (T—Fig. 134) on pump flange and timing gear housing. If pump timing marks cannot be properly aligned by shifting pump on mounting studs, the timing gear cover must be removed and gears retimed as outlined in paragraph 88.

COOLING SYSTEM

RADIATOR

All Models

130. To remove radiator, first drain cooling system and remove hood, side panels and grille. Remove air inlet tube and the cap screws attaching fan shroud to radiator, then move shroud back toward engine. Disconnect upper and lower radiator hoses and remove cap screws securing radiator to support frame, then lift radiator straight up out of frame assembly.

Install by reversing the removal procedure. Refill with a 50/50 mixture of water and ethylene glycol antifreeze.

THERMOSTAT

All Models

131. The thermostat is located in the water outlet housing attached to cylinder head. On MF255 model with AD4.203 engine the thermostat should start to open at 75°-78° C (167°-173° F) and should be fully open at 92° C (198° F). On all other models, the thermostat should begin to open at 80°-84° C (177°-183° F) and be completely open at 98° C (208° F).

WATER PUMP

All Models

132. R&R AND OVERHAUL. To remove water pump, first drain cooling system and remove hood and side guards. Remove alternator adjusting brackets. Remove fan and fan belt. Disconnect hoses from water pump. Remove pump mounting cap screws and withdraw water pump assembly from engine.

Refer to Fig. 140, 141 or 142 for an exploded view of water pump and thermostat housing. To disassemble the removed pump, remove nut (14), then pull the hub and pulley (13) from shaft. Notice and identify any screws that must be installed in mounting holes before pulley is reinstalled. Remove impeller housing or plate (1), then press shaft (4) and impeller (3) out toward rear of housing (6). Remove ceramic

counterface and seal (5). Remove snap ring (12), then press bearings (10) and spacer (11) out front of housing.

To assemble, press one bearing (10) onto shaft (4) with shielded side of bearing toward rear. Install spacer (11) then press other bearing (10) onto shaft with shielded side toward front. Pack both bearings and the cavity between bearings ½ full of high temperature lithium grease. Position flange (9) on shaft with

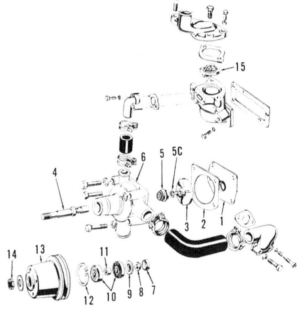

Fig. 140—Exploded view of water pump and thermostat used on MF255 tractors equipped with AD4.203 diesel engine.

1. Plate
2. Gasket
3. Impeller
4. Shaft
5. Seal
5C. Ceramic counterface
6. Shaft housing
7. Seal housing
8. Felt seal
9. Retainer
10. Bearings
11. Spacer
12. Snap ring
13. Pulley & hub
14. Nut
15. Thermostat

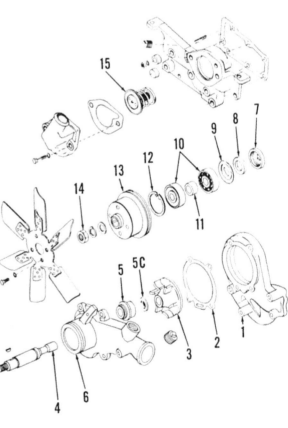

Fig. 141—Exploded view of typical water pump and thermostat used on tractors equipped with A4.236 and A4.248 diesel engines.

1. Impeller housing
2. Gasket
3. Impeller
4. Shaft
5. Seal
5C. Ceramic counterface
6. Shaft housing
7. Seal retainer
8. Felt seal
9. Flange
10. Bearings
11. Spacer
12. Snap ring
13. Pulley & hub
14. Nut
15. Thermostat

cupped side toward bearings (10), then position felt seal (8) and retainer (7) on shaft against flange. Press shaft, bearings, flange and felt seal assembly into housing (6) from the front until it bottoms. Install snap ring (12).

Make sure that cap screws that will be covered by pulley and hub are placed into holes in pump housing before installing pulley. Support rear of shaft (4), install key in shaft, then press hub and

pulley (13) onto shaft until bottomed. Install washer and retaining nut and tighten to 75 N·m (55 ft.-lbs.) torque.

Coat brass portion of rear seal (5) lightly with sealing compound and push it fully in place with carbon face toward rear. Install ceramic counterface (5C) with grooved side out toward impeller. Check shaft for freedom of rotation, then press impeller (3) onto rear of shaft until clearance (C—Fig. 143) between

impeller vanes and face of housing is 0.38-0.64 mm (0.015-0.025 inch) for MF255 models with AD4.203 diesel engine; 0.69-0.89 mm (0.027-0.035 inch) for all other models.

Use sealing compound on gasket and assemble impeller housing to shaft housing.

To reinstall pump, reverse the disassembly procedure.

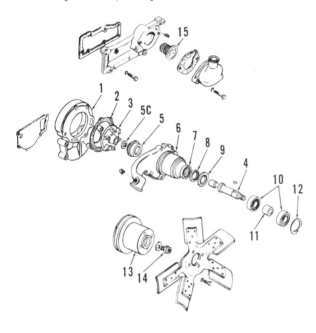

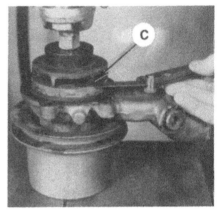

Fig. 142—Exploded view of water pump and thermostat housing used on MF255 and MF265 gasoline tractors. Refer to Fig. 141 for legend.

Fig. 143—Measure clearance (C) between impeller and housing while pressing impeller onto shaft. Refer to text.

IGNITION AND ELECTRICAL SYSTEM

DISTRIBUTOR

All Gasoline Models

135. TIMING. Timing marks are located on the crankshaft pulley as shown in Fig. 145. Initial (static) timing should be 12° BTDC for AG4.212 engine used in MF255 models; 11° BTDC for AG4.236 used in MF265 models. Maximum advance should be 36 crankshaft degrees BTDC above 2000 rpm for

Fig. 145—View of indicator and degree marks on crankshaft pulley on gasoline engines.

MF255 models; 32 crankshaft degrees BTDC above 2100 rpm for MF265 models. Timing should be checked with engine operating above maximum advance rpm. Turning distributor clockwise will advance timing. Firing order is 1-3-4-2 and distributor shaft rotates counterclockwise as viewed from top. Advance mechanism begins to operate at 900 crankshaft rpm for MF255; 500 crankshaft rpm for MF265 models. Delco-Remy 1112458 distributor is used on MF255 models; Delco-Remy 1112693 distributor is used on MF265 models.

136. OVERHAUL. Refer to Fig. 146 for an exploded view of the distributor. Centrifugal advance mechanism can be checked for binding or broken springs by turning rotor (21) counterclockwise and releasing after removing cap (23). Bushings are not available for housing (5); renew housing and/or shaft (8) if clearance is excessive. Shims (3) are available in thicknesses of 0.13 mm (0.005 inch) and 0.25 mm (0.010 inch)

for adjusting shaft end play which should be 0.05-0.25 mm (0.002-0.010 inch). Test specifications are as follows:

Delco-Remy 1112458
Breaker contact gap . . . . . . .0.56 mm
(0.022 in.)
Breaker arm spring
tension (measured at
center of contact . . . . . .5.3-6.4 N
(19-23 oz.)
Cam angle (degrees) 31-34

Advance data is in distributor degrees and distributor rpm.

Start advance . . . . .0-1 at 450
Maximum advance . . . . .11-13 at 1000

Delco-Remy 1112693
Breaker contact gap . . . . . . .0.56 mm
(0.022 in.)
Breaker arm spring
tension (Measured at
center of contact) 5.3-6.4
(19-23 oz.)
Cam angle (degrees) 31-34

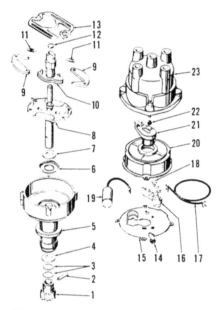

Fig. 146—Exploded view of ignition distributor showing component parts.

1. Drive gear
2. Roll pin
3. Shims
4. Thrust washer
5. Distributor housing
6. Seal
7. Thrust washer
8. Shaft assy.
9. Advance weight
10. Cam assy.
11. Spring
12. Oiler wick
13. Hold down plate
14. Insulator
15. Breaker plate
16. Point set
17. Primary wire
18. Eccentric screw
19. Condenser
20. Dust shield
21. Rotor
22. Brush
23. Distributor cap

Advance data is in distributor degrees and distributor rpm.

Start advance . . . . 0-1 at 250
Maximum advance . . . . . 10-12 at 1050

ALTERNATOR AND REGULATOR

137. A Delco-Remy 10-SI type alternator is used on some models, while other models may be equipped with either a Lucas or Perkins alternator. All alternators are equipped with a solid-state, nonadjustable voltage regulator.

Before any electrical system service is performed, a thorough check of condition of battery, cable connections, alternator belt and pulley should be made. Always disconnect battery leads before performing any service operations on electrical system. Do not connect or disconnect any part of charging system while the engine is running as damage to voltage regulator could result. When connecting battery leads, booster battery or battery charger, be sure to observe correct polarity (positive to positive, negative to negative). Do not short across or ground any terminals on alternator. Do not attempt to polarize alternator.

Refer to appropriate Figs. 310 through 312 for tractor wiring diagrams.

All Models with Delco-Remy Alternator

138. TESTING. Both 37 amp and 42 amp alternators are available, so be sure of alternator model when testing. Refer to the following specifications:

Delco-Remy No. 1100578
Field Current @ 27° C (80° F)
　Amperes 4.0-4.5
　Volts 12
Cold Output
　Amperes @ 2000 rpm 22
　Amperes @ 5000 rpm 33
Rated Hot Output
　(Amperes) @ Maximum
　Operating Speed 37

Delco-Remy No. 1100583
Field Current @ 27° C (80° F)
　Amperes 4.0-4.5
　Volts 12
Cold Output
　Amperes @ 2000 rpm 25
　Amperes @ 5000 rpm 37
Rated Hot Output
　(Amperes) @ Maximum
　Operating Speed 42

To test alternator output while installed on the tractor, first disconnect battery ground cable. Connect a test ammeter and voltmeter in circuit at alternator "BAT" terminal as shown in Fig. 147. Connect a carbon pile across battery as shown and turn on headlights. Operate engine at moderate speed and adjust carbon pile as required to obtain maximum current output. Voltage should be approximately 14.5 volts.

If ampere output is within 10 amperes of rated output stamped on alternator frame, alternator is operating satisfactorily. If output is not within 10 amperes of rated output, ground the field winding by inserting a screwdriver into test hole in rear of alternator (Fig. 148). If output is now within 10 amperes of rated output, renew regulator and

check field (rotor) winding as outlined in paragraph 139. If output is still not within 10 amperes of rated output, disassemble alternator and check field winding, diode trio, rectifier bridge and stator as outlined in paragraph 139.

If alternator fails to start charging at all, there may be an open circuit in wiring to No. 1 terminal of alternator. Voltage must be present at this terminal when key switch is turned to "RUN" position on gasoline models or when oil pressure is sufficient to close oil pressure switch on diesel models to provide initial excitation of field winding to start alternator charging. A quick field check for an open excitation circuit is to momentarily place a jumper wire between alternator "BAT" terminal and No. 1 terminal with engine running. If alternator starts charging, when it would not without jumper, an open excitation diode (located in wiring harness), oil pressure switch (diesel models) or ignition switch (gasoline models) is indicated.

139. OVERHAUL. To disassemble the alternator, first scribe matching marks (M—Fig. 149) on the two frame halves (4 and 16), then remove the four through-bolts. Pry frame apart with a screwdriver between stator frame (12) and drive end frame (4). Stator assembly (12) must remain with slip ring end frame (16) when unit is separated.

NOTE: When frames are separated, brushes will contact rotor shaft at bearing area. Brushes MUST be cleaned of lubricant if they are to be reused.

Clamp the iron rotor (13) in a protected vise, only tight enough to permit loosening of pulley nut (1). Rotor end frame can be separated after pulley and fan are removed. Check bearing surface of rotor shaft for visible wear or scoring. Examine slip ring surface for scoring or wear, and rotor winding for overheating or other damage. Check

Fig. 147—Connect a test ammeter and voltmeter in charging circuit as shown to check alternator output on tractor.

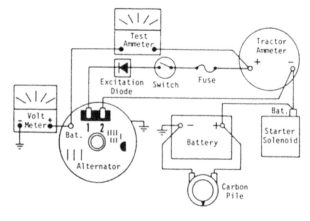

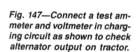

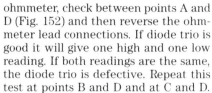

Fig. 148—View of rear of Delco-Remy alternator showing location of test hole and terminals.

rotor for grounded, shorted or open circuits using an ohmmeter as follows:

Refer to Fig. 150 and touch the ohmmeter probes to points (1-2) and (1-3); a reading near zero will indicate a short circuit to ground. Touch ohmmeter probes to the slip rings (2-3); reading should be 5.3-5.9 ohms. A higher reading will indicate an open circuit and a lower reading will indicate an internal short. If windings are satisfactory, mount rotor in a lathe and check runout at slip rings using a dial indicator. Runout should not exceed 0.05 mm (0.002 inch). Slip ring surface can be trued if runout is excessive or if surfaces are scored. Finish with 400 grit or finer polishing cloth until scratches or machine marks are removed.

Before removing stator, brushes or diode trio, refer to Fig. 151 and check

for grounds between the two insulated screws and regulator ground screw with an ohmmeter, using the lowest range scale. Then reverse the lead connections. If both readings are the same when tester leads are reversed, the brushes may be grounded because of defective insulating washer and sleeve at the two screws. If the screw assembly is not damaged or grounded, the regulator is defective.

To test the diode trio, first remove the stator. Then remove the diode trio, noting the insulator positions. With an

ohmmeter, check between points A and D (Fig. 152) and then reverse the ohmmeter lead connections. If diode trio is good it will give one high and one low reading. If both readings are the same, the diode trio is defective. Repeat this test at points B and D and at C and D.

The rectifier bridge (Fig. 153) has a grounded heat sink (A) and an insulated heat sink (E) that is connected to the output terminal. Connect ohmmeter to the grounded heat sink (A) and to the flat metal strip (B). Then reverse the ohmmeter lead connections. If both readings are the same, the rectifier bridge is defective. Repeat this test between A and C, A and D, B and E, C and E, and D and E. Capacitor (11—Fig. 149) connects to the rectifier bridge and

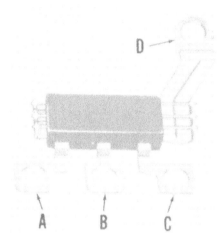

Fig. 152—Diode trio test points. Refer to text.

Fig. 150—Removed rotor assembly showing test points to be used when checking for grounds, shorts and opens.

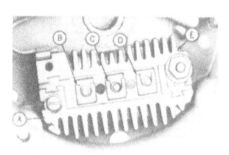

Fig. 153—Bridge rectifier test points. Refer to text.

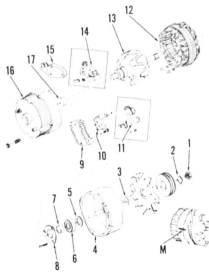

Fig. 149—Exploded view of Delco-Remy alternator used on some models. Scribe alignment marks (M) on end frame prior to disassembly.

1. Pulley nut
2. Washer
3. Spacer (outside drive end)
4. Drive end frame
5. Grease slinger
6. Ball bearing
7. Spacer (inside drive end)
8. Bearing retainer
9. Bridge rectifier
10. Diode trio
11. Capacitor
12. Stator
13. Rotor
14. Brush holder
15. Solid state regulator
16. Slip ring end frame
17. Bearing & seal assy.

Fig. 151—View of slip ring end frame and related components.

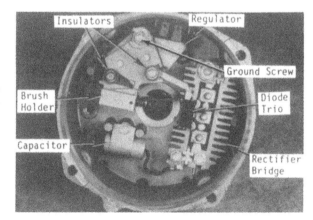

49

grounds to end frame, and protects the diodes from voltage surges.

Test the stator windings for grounded or open circuits as follows: Connect ohmmeter leads successively between each pair of leads. A high reading will indicate an open circuit. The three stator leads have a common connection in the center of the windings. Connect ohmmeter leads between each stator lead and stator frame. A very low reading would indicate a grounded circuit. A short circuit within the stator windings cannot be readily determined by test because of the low resistance of the windings.

Brushes and springs are available only as an assembly which includes brush holder (14—Fig. 149). If brushes are reused, make sure all grease is removed from surface of brushes before unit is reassembled. When reassembling, first install regulator and then brush holder, springs and brushes. Push brushes up against spring pressure and insert a short piece of straight wire through hole in brush holder and through end frame to outside. Be sure that the two screws at points A and B (Fig. 151) have insulating washers and sleeves.

NOTE: A ground at these points will cause no output, or controlled output.

Remove and inspect ball bearing (6—Fig. 149). If bearing is in satisfactory condition, fill bearing ¼ full with Delco-Remy lubricant No. 1948791 and reinstall. Inspect needle bearing (17) in slip ring end frame. This bearing should be renewed if its lubricant supply is exhausted; no attempt should be made to relubricate and reuse the bearing. Press old bearing out toward inside and press new bearing in from outside until flush with outside of end frame. Saturate felt seal with SAE 20 oil and install seal.

Reassemble alternator by reversing the disassembly procedure. Be sure insulators are installed at points shown in Fig. 151. Remove wire retaining the brushes after end frames are bolted together. Tighten pulley retaining nut to a torque of 68 N·m (50 ft.-lbs.).

All Models With Lucas Alternator

140. TESTING. To check alternator wiring harness continuity, disconnect wiring connections from alternator and turn start switch to auxiliary position. Connect a voltmeter between ground and each of the disconnected leads. Battery voltage should be present at the alternator ''+'' (battery) lead and ''IND'' (warning light) lead. A zero reading indicates an open circuit in wiring or faulty warning light bulb. Check

continuity of alternator ground lead with an ohmmeter.

If voltage is present at ''+'' and ''IND'' leads, but warning light fails to come on when leads are reconnected to alternator, a faulty regulator and/or field (rotor) winding is indicated.

To check alternator output, first disconnect battery ground cable. Disconnect wiring from alternator, remove cover from rear of alternator, then reconnect wiring leads. Connect an ammeter between starter solenoid terminal and alternator terminal (Fig. 154). Connect a voltmeter between output terminal and ground. Connect a carbon pile across battery terminals. Reconnect battery ground cable and start engine. Operate engine at 2000 rpm, turn on headlights and adjust carbon pile as required to obtain full current output. Ammeter reading should be approx-

imately 34 amperes. Adjust carbon pile so ammeter reading is less than 10 amperes. Voltage reading should be within range of 13.6-14.4 volts.

If output is low, use a jumper lead to short together the ''F'' and ''-'' connections (Fig. 154) on alternator. If output is now within specified range, a faulty regulator and/or field winding is indicated. If output is still low, fault is elsewhere in alternator. Disassemble and check as outlined in paragraph 141.

141. OVERHAUL. To disassemble alternator, first remove rear cover (17—Fig. 155). Disconnect and remove surge protection diode (15), regulator (16) and brush holder (13) with brushes (12). Note that these parts can be removed without removing the alternator from the engine. Any further disassembly will require removal of alternator.

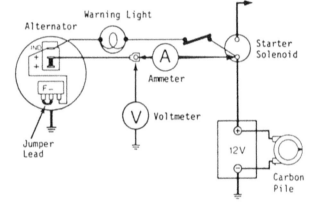

Fig. 154—Connect ammeter and voltmeter to charging circuit as shown to check output of Lucas alternator. Refer to text.

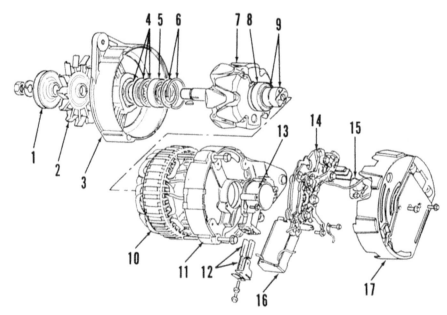

Fig. 155—Exploded view of Lucas alternator assembly used on some models.

1. Pulley	5. Bearing	9. Slip rings	14. Rectifier
2. Fan	6. Retaining ring & plate	10. Stator	15. Surge protection diode
3. Drive end frame	7. Rotor	11. Slip ring end frame	16. Regulator
4. Felt seal & retainer assy.	8. Bearing	12. Brushes & springs	17. Rear cover
		13. Brush holder	

Scribe match marks on slip ring housing (11), stator frame (10) and drive end frame (3) to ensure correct alignment for reassembly. Unsolder stator wires from rectifier (14), then remove rectifier from housing. Remove retaining bolts and withdraw housing and stator from end frame.

Remove pulley and fan from rotor shaft, then press rotor assembly (7) out of front bearing (5) and end frame (3). Remove snap ring and retainer plate (6), bearing (5) and felt seal and retainer assembly (4) from end frame. To remove rear bearing (8), the slip ring (9) connections must first be unsoldered and the rings removed from shaft. Use a suitable puller to remove the bearing.

Inspect brushes and rotor slip rings for wear and renew if necessary. Brush length when new is 20 mm (0.787 inch). Renew brushes if length is less than 10 mm (0.394 inch). Brush spring tension should be 1.3-2.7 N (4.7-9.8 ounces).

Connect ohmmeter leads between the two rotor slip rings to check field winding resistance. Ohmmeter should register 3.2 ohms. Check for continuity between slip rings and rotor frame. If continuity is indicated, field winding is grounded and rotor should be renewed.

Check directional continuity of rectifier diodes using a battery powered 12 volt test light or an ohmmeter. Touch one test lead to diode stem and other test lead to rectifier plate, then reverse the test lead connections. There should be continuity in one direction only. If there is continuity in both directions or if there is no continuity in either direction, the diode is defective. If any diode checks faulty, rectifier assembly must be renewed.

Test stator windings for grounded or open circuits as follows: Connect ohmmeter leads between stator frame and each lead wire. A low reading would indicate a grounded circuit. Connect tester leads successively between each pair of stator leads. A high reading indicates an open circuit.

Reassemble alternator by reversing the disassembly procedure. Be sure to align match marks on housings made prior to disassembly. Tighten pulley retaining nut to 38 N·m (28 ft.-lbs.) torque.

All Models With Perkins Alternator

142. TESTING. To check alternator output, connect a voltmeter between "D+" terminal of alternator and ground. Operate engine at moderate speed and observe voltmeter reading. Voltage should be within range of 13 to 15 volts.

If voltmeter reading exceeds 15.5 volts, voltage regulator is faulty. If voltmeter reading is less than 13 volts, insert a pin into small hole near the center of alternator rear housing to ground the "F" terminal. If voltage is now within range of 13 to 15 volts, regulator is faulty. If voltage remains below 13 volts, disassemble alternator and check field winding, diodes and stator as outlined in paragraph 143.

143. OVERHAUL. Prior to disassembly, scribe matching marks across end frame (4—Fig. 157) and rear housing (14) to ensure correct alignment for reassembly. Remove four through-bolts, then separate end frame with rotor from rear housing and stator. Hold rotor in a vise, then remove pulley nut, pulley (1), fan (2) and spacer (3) from rotor shaft. Separate rotor from end frame. Remove front bearing (5) from end frame and rear bearing (8) from rotor if necessary.

Remove five nuts attaching brush holder/diode/stator assembly to rear housing. Unsolder stator wires from diodes. Drill out rivet between brush holder (12) and diode rectifier (11). Unsolder brush box and regulator from rectifier. Unsolder the regulator to brush holder terminals and separate regulator from brush holder.

Renew brush holder assembly if brush "free" length from end of holder is less than 2 mm (0.080 inch). This minimum allowable length is marked on the brushes by a groove on one side of each brush.

Current flows in one direction only in each of the six diodes. Check each diode for directional continuity using a battery powered 12 volt test lamp or an ohmmeter. Connect tester leads to positive and negative side of diode, then reverse tester lead connections. Test lamp should light or ohmmeter should indicate continuity in one direction only. If continuity is indicated in both directions or if there is no continuity in either direction, diode is defective and rectifier assembly must be renewed.

Check rotor field winding for open circuit by connecting test lamp or ohmmeter leads to each of the rotor slip rings. If no continuity is indicated, rotor is faulty and must be renewed. Check for continuity between the slip rings and rotor shaft. If continuity is indicated, winding is grounded and rotor must be renewed.

Check for continuity between each of the stator coil leads. Renew stator if an open circuit is indicated. Check for continuity between each of the stator coil leads and the stator frame. If continuity is indicated, winding is grounded and stator must be renewed.

To reassemble alternator, reverse the disassembly procedure while noting the following special instructions: Tighten brush holder/diode assembly retaining nuts to 3.1-3.9 N·m (28-34 in.-lbs.) torque. Tighten pulley retaining nut to 44-59 N·m (33-43 ft.-lbs.) torque. Push brushes up into holder and insert a pin through hole in center of rear cover to

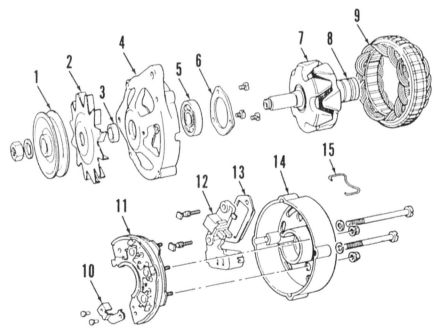

Fig. 157—Exploded view of Perkins alternator used on some models.

1. Pulley	6. Retainer plate	10. Diode terminal
2. Fan	7. Rotor	11. Diode rectifier	13. Regulator
3. Spacer	8. Bearing	12. Brushes & holder	14. Slip ring end frame
4. Drive end frame	9. Stator	assy.	15. Lock clip
5. Bearing		

hold brushes in this position. Remove brush retaining pin after rotor is installed. Tighten the through-bolts to a torque of 3.1-3.9 N·m (28-34 in.-lbs.).

STARTING MOTORS

Early tractor models were equipped with Delco-Remy starting motors, while later production tractors use Lucas or Perkins (Hitachi) starting motors.

All Models with Delco-Remy Starter

144. Several different models of Delco-Remy starting motors have been used. Be sure to note Delco-Remy model number stamped on starter name plate when servicing Delco-Remy starting motor. Refer to the following specifications:

D-R Model No. 1108396
No-Load Test:
Volts 9
Amperes (w/solenoid) 50-80
Rpm 5500-10500
Solenoid Current Consumption:
Pull-in Winding—
Amps 13-15.5
Volts 5
Hold-in Winding—
Amps 14.5-16.5
Volts .10

D-R Model No. 1107870 & 1107872
No-Load Test:
Volts 9
Amperes (w/solenoid) 50-80
Rpm 5500-9000
Solenoid Current Consumption:
Pull-in Winding—
Amps 13-15.5
Volts 5
Hold-in Winding—
Amps 14.5-16.5
Volts .10

D-R Model No. 1109253 & 1109257
No-Load Test:
Volts 9
Amperes (w/solenoid 20-120
Rpm 9000-14000
Solenoid Current Consumption:
Pull-in Winding—
Amps 13-15.5
Volts 5
Hold-in Winding—
Amps 14.5-16.5
Volts .10

D-R Models No. 1113650
No-Load Test:
Volts 9
Amperes (w/solenoid) 75-105
Rpm 5000-7000

Solenoid Current Consumption:
Pull-in Winding—
Amps 26-29
Volts 5
Hold-in Winding—
Amps 18-20
Volts .10

Refer to Fig. 159 for exploded view of typical Delco-Remy starting motor. Starter drive pinion clearance is not adjustable; however, some clearance must be maintained between end of pinion and starter drive frame to assure solid contact of the magnetic switch. Normal pinion clearance should be within the limits of 0.25-3.56 mm (0.010-0.140 inch). Connect a 6 volt battery to solenoid terminals to actuate solenoid but keep armature from turning when measuring pinion clearance.

All Models With Lucas Starter

145. A Lucas starting motor is used on some tractors. Connect voltmeter and ammeter to starting motor as shown in Fig. 158 to perform no-load test. Specifications are as follows:

Lucas Model No. M113
No-Load Test—
Volts 11
Amperes (max.) 100
Rpm 5000-7500

Lucas Model No. M127
No-Load Test—
Volts 11
Amperes (min.) 115
Rpm 5500-8000

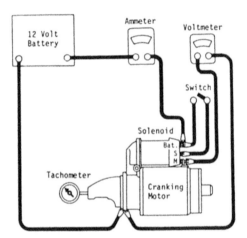

Fig. 158—Connect a voltmeter and ammeter to starting motor as shown to perform starter no-load test. Refer to text.

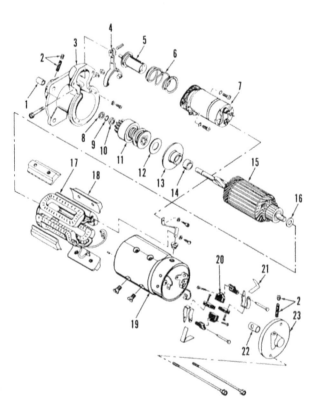

Fig. 159—Exploded view of typical Delco-Remy starting motor used on some tractors. Items 12, 13 and 14 are not used on some models.

1. Bushing
2. Oil wick & plug
3. Drive end housing
4. Shift lever
5. Plunger
6. Spring
7. Solenoid switch
8. Thrust washer
9. Snap ring
10. Pinion stop collar
11. Pinion drive assy.
12. Fiber washer
13. Center bearing plate
14. Bushing
15. Armature
16. Washer
17. Field coil
18. Pole shoe
19. Field frame
20. Brushes
21. Brush spring
22. Bushing
23. End cover

Refer to Fig. 160 for exploded view of typical Lucas starting motor. Brush spring tension should be between 11.6-15.5 N (42-55 ounces). Renew brushes if length is less than 8 mm (5/16 inch). When renewing armature bushings, new bushings should be soaked in engine oil for 24 hours prior to installation. Tighten the two through-bolts to a torque of 11 N·m (8 ft.-lbs.).

Starter drive pinion end clearance is adjustable. Connect a 6 volt battery to blade terminal on solenoid to hold pinion in engaged position. While pushing pinion lightly toward armature to remove any free play in linkage, measure clearance between end of pinion (13) and the thrust collar (14). Turn eccentric pin (16) as required to obtain recommended clearance of 0.13-0.38 mm (0.005-0.015 inch).Tighten locknut to secure the pin.

All Models With Perkins Starter

146. A Perkins starting motor is used on some models. Connect voltmeter and ammeter to starting motor as shown in Fig. 158 to perform no-load test. Specifications are as follows:

No-Load Test—
Volts 12
Amperes (max.) 120
Rpm 4000-7500

Refer to Fig. 161 for an exploded view of typical Perkins starting motor. Armature commutator minimum allowable diameter is 40 mm (1.575 inches). Maximum allowable runout for armature is 0.10 mm (0.004 inch). Depth of insulation below surface of commutator should be 0.20-0.80 mm (0.008-0.031 inch). Minimum allowable brush length is 14 mm (0.551 inch).

Armature end play should be 0.05-0.30 mm (0.002-0.012 inch). If end play exceeds 0.30 mm (0.012 inch), install an additional thrust washer (15).

Starter drive pinion end clearance is adjustable. Connect a 6 volt battery to solenoid "S" terminal to hold pinion in engaged position while measuring clearance. Push pinion lightly toward armature to remove free play in linkage, then measure clearance between face of pinion gear and the pinion stop (19).

Recommended clearance is 0.3-1.5 mm (0.012-0.060 inch). Add or remove space washers (2) as required to obtain desired clearance.

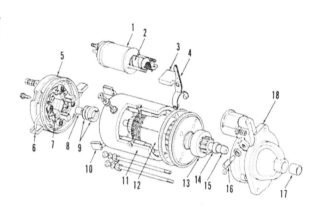

Fig. 160—Exploded view of Lucas starting motor used on some tractors.
1. Solenoid assy.
2. Plunger
3. Seal
4. Shift lever
5. End cover
6. Brushes
7. Brush plate assy.
8. Bushing
9. Brake assy.
10. Brushes
11. Field frame
12. Armature
13. Drive pinion assy.
14. Thrust collar
15. Snap ring
16. Eccentric pin
17. Bushing
18. Drive end frame

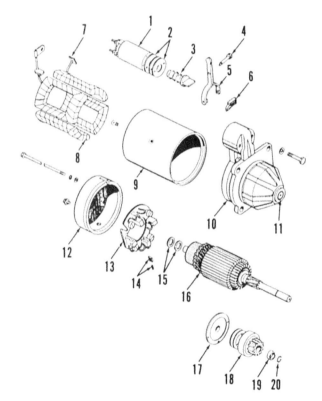

Fig. 161—Exploded view of Perkins starting motor used on some tractors.
1. Solenoid
2. Spacer washers
3. Plunger
4. Pivot pin
5. Shift lever
6. Dust seal
7. Brush
8. Field coil
9. Field housing
10. Drive end frame
11. Bushing
12. End cover
13. Brush holder assy.
14. Brush & spring
15. Washers
16. Armature
17. Center plate
18. Drive pinion assy.
19. Pinion stop collar
20. Snap ring

ENGINE CLUTCH

Tractors may be equipped with a flywheel mounted dual stage clutch which permits continuous (live) pto or with a split torque clutch with independent pto. Refer to the appropriate following paragraphs for adjustment and overhaul procedures.

ADJUSTMENT
All Models

147. Clutch pedal free play, measured between clutch release arm and transmission housing as shown at (F—Fig. 165) should be 4 mm (5/32 inch). To

adjust, loosen release arm clamp bolt (1), insert a punch through hole in end of release shaft (2) and rotate shaft clockwise until throwout bearing just contacts clutch fingers. Move pedal until clearance (F) is correct, then tighten clamp bolt.

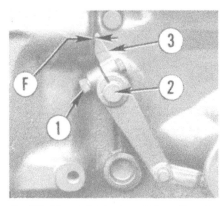

Fig. 165—View of clutch release linkage. Free play (F), measured between release arm (3) and transmission housing, should be 4 mm (5/32 inch). Free play is adjusted by changing the installed position of released arm on cross shaft (2).

TRACTOR FRONT SPLIT

All Models

148. To separate engine from transmission, remove instrument panel covers, grille and hood. Disconnect battery cables and wiring to rear lights. Remove the starter motor. Remove cap screws attaching steering column lower support to transmission housing. Disconnect Multi-Power shift linkage (if equipped), oil cooler lines at top of transmission, exhaust pipe (low exhaust models only) and foot throttle linkage (if equipped).

Install wedge blocks between front axle and axle support to prevent tipping. Remove front mounted weights if so equipped. Support engine and transmission separately with suitable splitting stands. Remove cap screws attaching engine to transmission, then carefully move engine and front end assembly forward from transmission housing.

Install a guide stud in each side of transmission housing to assist in alignment when rejoining tractor. Be sure that engine and transmission flanges meet before tightening attaching cap screws. DO NOT use the cap screws to force the two units together. Tighten attaching cap screws to 75 N·m (55 ft.-lbs.) torque.

DUAL CLUTCH UNIT

All Models So Equipped

149. REMOVE AND REINSTALL. To remove clutch assembly, first separate engine from transmission as outlined in paragraph 148. Punch mark the clutch cover (16—Fig. 168 or 169), pressure plate (14), drive plate (12), pressure plate (7) and engine flywheel prior to removing clutch so components can be

reinstalled in their original positions if reused. Install the three special "T" bolts shown in Figs. 166 and 167 and tighten the forcing nuts to compress clutch springs. Remove the retaining cap screws and lift clutch assembly from flywheel.

It is recommended that guide studs be used in flywheel when reinstalling clutch. Position inner clutch disc (6—Fig. 168 or 169) with hub facing away

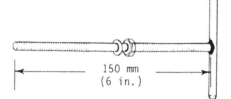

Fig. 166—Special "T" bolts can be made to facilitate clutch removal, repair and installation. A cross handle should be welded to a 75 mm (6 inch) piece of ¼ inch-20 threaded rod. A forcing nut and washer are used to compress the clutch springs.

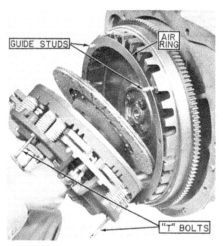

Fig. 167—View showing removal/installation of typical dual clutch. Air ring is not used on some models.

from flywheel. On models so equipped, position spacer ring (5—Fig. 168) on flywheel. On all models, use a suitable pilot shaft (such as MFN753) through hubs of clutch discs to align splines. Install clutch assembly aligning previously made punch marks. Remove "T" bolts and adjust clutch levers as outlined in paragraph 151.

150. OVERHAUL. Prior to disassembling clutch, punch mark clutch cover (16—Fig. 168 or Fig. 169), pressure plate (14), drive plate (12) and pressure plate (7) so components can be reassembled in their original positions if reused.

To disassemble clutch unit shown in Fig. 168, unhook the three tension springs (22) from release levers (17). Loosen the forcing nuts on the three "T" bolts until it is possible to drive the retaining pins (21), if so equipped, out of the release lever pivot pins (20). Do not attempt to drive the retaining pins completely out of clutch cover while clutch is assembled as Belleville spring (15) may be damaged. Pins can be removed from cover after clutch is disassembled. On later style clutches, a hairpin type retainer clip is used to retain pivot pins (20). Remove release lever pivot pins, then loosen forcing nuts on "T" bolts until clutch spring pressure is relieved. Remove "T" bolts and separate components of clutch.

To disassemble clutch unit shown in Fig. 169, remove pivot pin retainer clips (21). Drive out release lever pivot pins (20) and remove lever torsion springs (22). Loosen forcing nuts on "T" bolts to relieve spring tension. Remove "T" bolts and separate components of clutch.

On all models, examine all parts and renew any which are excessively worn or damaged. Flywheel friction face may be resurfaced up to a maximum depth of 1.0 mm (0.040 inch). Note that same amount of material must also be removed from clutch mounting surface on flywheel to maintain original distance between clutch mounting surface and

Fig. 168—Exploded view of dual clutch assembly used on Models MF255, MF265 and MF270. A hairpin type retainer clip is used in place of pin (21) on later style clutches to retain pivot pins (20).

1. Pin
2. Cotter pin
3. Link pin
4. Link
5. Spacer ring
6. Transmission disc
7. Pressure plate
8. Locknut
9. Insulating washer
10. Clutch spring
11. Adjusting screw
12. Drive plate
13. Pto disc
14. Pressure plate
15. Belleville spring
16. Clutch cover
17. Release lever
18. Locknut
19. Adjusting screw
20. Pivot pin
21. Retainer pin
22. Torsion spring

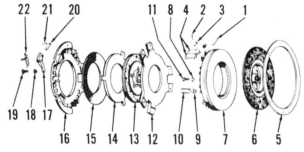

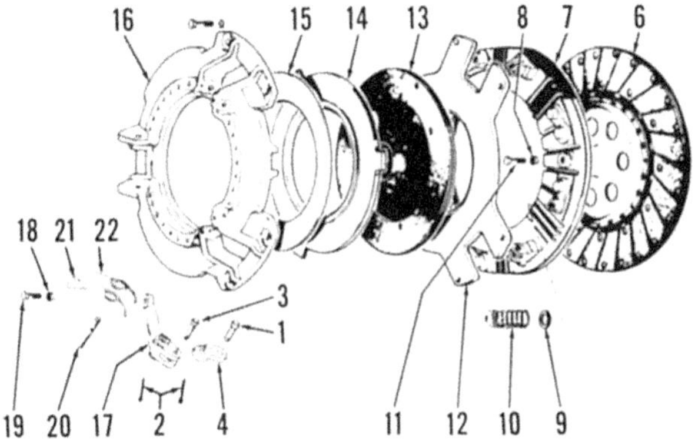

Fig. 169—Exploded view of dual clutch assembly used on Models MF275 and MF290. Refer to Fig. 168 for legend.

SPLIT TORQUE CLUTCH

Models MF255 and MF265 With Early Style Clutch

152. REMOVE AND REINSTALL. Refer to paragraph 148 and separate engine from transmission. Punch mark the clutch cover, pressure plate and flywheel so that original balance can be maintained when reassembling. Loosen pressure plate retaining screws evenly to prevent warpage.

Install clutch driven disc with side marked "FLYWHEEL SIDE" toward flywheel. The center hub will be riveted to rear side of clutch disc center. Use a suitable aligning tool (such as MFN753) to align driven disc and pto drive splines in cover. Align the previously affixed punch marks and tighten the retaining screws evenly to a final torque 34- 52 N·m (25-38 ft.-lbs.).

Use the 11 inch side of Massey-Ferguson tool MFN202D and adjust release lever height. Head of release lever adjusting screw should just touch gage as shown in Fig. 174.

NOTE: Release lever height should be set with new clutch disc installed, even if old disc is to be reinstalled.

friction face of flywheel. Original distance between mounting surface and friction surface is 43.63-43.75 mm (1.718-1.723 inches) on early style clutch used on Models MF255 and MF265, or 39.63- 39.75 mm (1.560-1.565 inches) on all other models.

CAUTION: It is not permissible to resurface drive plate (12) or either of the pressure plates (7 and 14).

It is recommended that clutch spring insulating washers (9) be renewed when reassembling clutch. To reassemble clutch, reverse the disassembly procedure making sure that assembly punch marks are aligned. Insert "T" bolts and tighten forcing nuts evenly to compress clutch springs and hold clutch components together.

Reinstall clutch on flywheel using a suitable pilot shaft to align clutch disc splines. Adjust clutch as outlined in paragraph 151 before rejoining engine to transmission.

151. ADJUSTMENT. After the clutch assembly has been installed in flywheel, two adjustments are necessary for proper clutch operation.

NOTE: A new transmission clutch disc (6—Fig. 168 or Fig. 169) must be installed to correctly adjust release lever height, even if old disc will be reused. Make the adjust-

ment with new disc installed, then remove clutch and reinstall worn disc without changing the adjustment.

To adjust clutch release levers to correct height, use special tool MFN202C for early style clutch used in MF255 and MF265 models or special tool 6179 for all other models. Make sure that head of release lever adjusting screws just touch the gage as shown in Fig. 171 or Fig. 172.

Adjust pto pressure plate adjusting screws to provide 2.0 mm (0.080 inch) clearance between head of screws and drive plate lugs as shown in Fig. 173.

Rejoin tractor as outlined in paragraph 148 and adjust clutch linkage as outlined in paragraph 147.

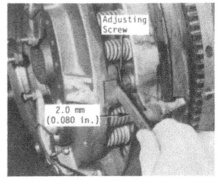

Fig. 173—Adjust pto pressure plate adjusting screws to provide 2.0 mm (0.080 inch) clearance between head of screws and drive plate lugs.

Fig. 170—View showing installation of pivot pin retainer clip on MF275 and MF290 dual clutch.

Fig. 172—Use special tool 6179 to adjust clutch release levers to correct height on MF275 and MF290 dual clutch.

Fig. 174—Use side of special tool (MFN 202D) marked 11 inches to adjust release lever height on split torque clutch used in MF255 and MF265 models.

Fig. 171—Use special tool MFN 202C to adjust clutch release levers to correct height on MF255, MF265 and MF270 dual clutch.

153. OVERHAUL. To disassemble the removed pressure plate and cover unit, place the assembly in a press as shown in Fig. 175 and apply only enough pressure to relieve tension on pins. Remove pins (7—Fig. 176) and lift off cover.

Inspect release levers and pins for wear or damage and pressure plate for scoring, heat checks or wear at actuating pin holes. Flywheel friction face may be resurfaced up to a maximum depth of 1.0 mm (0.040 inch). An equal amount of material must be removed from clutch mounting surface on flywheel to maintain original clutch mounting surface to friction surface dimension of 35.48-35.99 mm (1.397-1.417 inches).

CAUTION: It is not permissible to resurface the clutch pressure plate friction face.

Assemble by reversing the disassembly procedure, making certain that previously affixed marks on cover and pressure plate are aligned. Install pins with heads leading in normal direction of rotation as shown in Fig. 175. Clutch levers should be adjusted to equal height after installation using a new clutch disc as outlined in paragraph 152.

All Models With Late Style Clutch

154. REMOVE AND REINSTALL. Refer to paragraph 148 and separate engine from transmission. Punch mark the clutch cover, pressure plate and flywheel so that original balance can be maintained when reassembling. Install the three special bolts shown in Fig. 166 and tighten the forcing nuts to compress the clutch springs. Remove the retaining cap screws and lift clutch assembly from flywheel.

Use guide studs in flywheel and position disc (17—Fig. 177) with hub away from flywheel. Use pilot tool such as MFN753 through pto drive hub (12), clutch disc (17) and into pilot bearing. Install clutch assembly with previously affixed punch marks aligned. Tighten retaining screws to 27- 34 N·m (20-25 ft.-lbs.) torque, then remove "T" bolts.

Adjust release lever height using special tool 6179. Head of lever adjusting screws should just contact gage. Distance from inner lip of flywheel to top of lever adjusting screws should be 82.55-83.31 mm (3.250-3.280 inches).

NOTE: Release lever height should be set with a new clutch disc installed, even if old disc is to be reinstalled.

155. OVERHAUL. Be sure that cover plate (11—Fig. 177), pto drive hub (12) and pressure plate (16) are punch marked for correct reassembly before disassembling clutch. Clutch springs must be compressed using "T" bolts and forcing nuts or a press before release lever pivot pins can be removed. Pry retainer pins (2) out, then drive pivot pins (3) from release levers (5). Remove release lever torsion springs (4), then relieve clutch spring tension slowly and separate components of clutch.

Inspect all parts thoroughly and renew any which are damaged or worn. Renew clutch springs if distorted or discolored from heat. Renew insulating washers (15) when reassembling. Friction face of flywheel may be resurfaced up to a maximum depth of 1.0 mm (0.040 inch). An equal amount of material must be removed from clutch mounting surface on flywheel to maintain original distance of 39.62-39.75 mm (1.560-1.565 inches) from clutch mounting surface to friction surface.

CAUTION: It is not permissible to resurface pressure plate friction surface.

To reassemble clutch, reverse the disassembly procedure.

Fig. 175—Use a press to disassemble split torque clutch used on MF255 and MF265 models. Be sure unit is reassembled with head of pivot pin on leading side as shown.

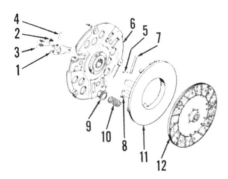

Fig. 176—Exploded view of MF255 and MF265 split torque clutch assembly.

1. Release lever
2. Adjusting screw
3. Torsion spring
4. Floating pin
5. Cotter pin
6. Clutch cover
7. Retaining pin
8. Washer
9. Spring cup
10. Spring
11. Pressure plate
12. Clutch disc

Fig. 177—Exploded view of typical split torque clutch used in MF270, MF275 and MF290 models.

1. Adjusting screw
2. Retainer pin
3. Pin
4. Torsion spring
5. Release lever
6. Pin
7. Cotter pin
8. Pin
9. Link
11. Cover plate
12. Drive plate
13. Bolt & nuts
14. Spring
15. Insulating washer
16. Pressure plate
17. Clutch disc

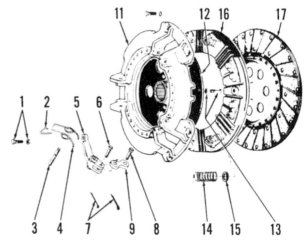

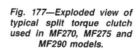

TRANSMISSION

All models may be equipped with either a gear change transmission having eight forward speeds and two reverse speeds or a "Multi-Power" transmission which has 12 speeds forward and four reverse speeds. The Multi-Power transmission is a gear change transmission with six forward and two reverse speeds which is additionally equipped with a hydraulically operated high-low range unit.

Many of the service procedures and some of the parts are similar for the 8-speed and Multi-Power transmissions.

Refer to the appropriate following paragraphs for service of both transmissions.

Main input shaft retainer, pto input shaft and Multi-Power shift control valve (if so equipped) can be serviced after separating engine from transmission as outlined in paragraph 148. High-low planetary unit, shift rails and forks can be serviced after separating transmission housing from rear axle and differential housing as outlined in paragraph 156. Transmission mainshaft with related gears and reverse idler gear can also be serviced after performing ''rear split'' if desired. Transmission assembly must be removed as outlined in paragraph 157 to service all other components.

TRACTOR REAR SPLIT

All Models

156. To separate tractor between transmission housing and differential housing, proceed as follows: Drain oil from transmission and differential housings. Remove step plates from both sides of tractor. Disconnect wiring to rear of tractor. Disconnect horizontal exhaust pipe if so equipped. Disconnect clutch linkage. Disconnect foot brake rods (mechanical brake models) or foot brake hydraulic lines (hydraulic brake models) and parking brake linkage. Remove clutch and brake pedals with cross shaft and hand brake lever (if equipped). If equipped with Multi-Power transmission, independent pto or auxiliary hydraulic system, remove transmission shift cover as outlined in paragraph 158 and disconnect hydraulic line inside transmission housing.

Install wedges between front axle and axle support to prevent tipping. Place suitable splitting stands and floor jack under transmission housing and differential housing. Remove bolts attaching transmission to differential housing, install guide studs and carefully separate tractor.

To rejoin tractor, shift high/low planetary into low range and position rear drive shaft into planetary coupler. Remove hitch response control cover from right-hand side of differential housing or remove pto control lever cover from left-hand side of differential housing so alignment of rear drive shaft coupling and pto lower drive shaft can be observed. Place a new gasket on guide studs, then carefully join transmission to differential housing.

NOTE: Make certain that housings are completely together before retaining bolts are tightened.

Check the clearance between rear drive shaft coupler tube and the locking collar on differential drive pinion with rear drive shaft pushed fully forward. Clearance should be 0.4-2.5 mm (0.015-0.100 inch). If necessary, reposition pin forward in coupler tube holes to reduce clearance or move pin rearward to increase clearance.

Tighten housing retaining bolts in a clockwise direction (as viewed from rear of tractor) starting at top center bolt. Tighten bolts in two steps to a final torque of 105 N·m (75 ft.-lbs.). Complete installation by reversing the removal procedure.

TRANSMISSION REMOVAL

All Models

157. To remove the complete transmission unit from the tractor, first drain transmission and hydraulic system fluid and separate engine from front of transmission housing as outlined in paragraph 148. If equipped with Multi-Power transmission or independent pto, remove transmission top (shift) cover and disconnect hydraulic line inside transmission housing. Disconnect the foot brake rods (mechanical brake models) or brake pipes (hydraulic brake models). Disconnect hand brake linkage and clutch pedal rod, then remove pedals with linkage and hand brake lever with linkage.

Block under the front of rear axle and differential housing and under drawbar to prevent tipping. Support transmission housing with a suitable floor jack or hoist. Remove bolts attaching transmission to differential housing, install guide studs, then slide transmission assembly forward from differential housing.

To reinstall transmission, first shift high/low planetary into low range and main transmission into third gear. Position rear drive shaft into planetary unit coupler. Remove hitch response control cover from right side of differential housing or remove pto control lever cover from left side of housing so alignment of rear drive shaft coupling and pto lower drive shaft can be observed. Place a new gasket on guide studs, then carefully join transmission to differential housing.

NOTE: Make certain that housings are completely together before retaining nuts and bolts are tightened.

Push the rear drive shaft and coupler forward until spring in front of drive shaft is fully compressed then measure the distance between rear of coupler tube and front of locking collar on differential drive pinion. Clearance should

be 0.4-2.5 mm (0.015-0.100 inch). If clearance is too large, reposition pin in holes in coupler tube closer to the front. Move pin rearward if clearance is too small.

Tighten retaining bolts in a clockwise direction (as viewed from rear of tractor) starting at top center bolt. Tighten bolts in two steps to a final torque of 105 N·m (75 ft.-lbs.). Rejoin transmission to engine and complete installation by reversing removal procedure.

TRANSMISSION TOP COVER

All Models

158. REMOVE AND REINSTALL. To remove transmission top (shift) cover, first remove instrument console lower panels. Shift transmission to neutral, then remove gear shift lever retaining pins (4—Fig. 180) and lift shift levers from cover. Remove cap screws attaching steering support and instrument console to shift cover. Remove cap screws securing shift cover to transmission housing. Work the shift cover from beneath console being careful not to drop detent springs into transmission housing.

Install top (shift) cover in reverse of removal procedure. Be sure that shift levers properly engage shift forks.

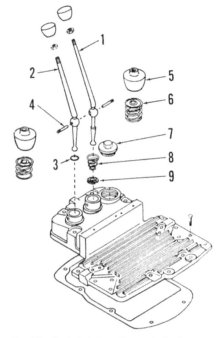

Fig. 180—Exploded view of transmission top cover and shift levers typical of all models.

1. Gearshift lever
2. High/Low shift lever
3. ''O'' ring
4. Pins (2)
5. Boot
6. Spring
7. Oil filler plug
8. Spring
9. Seat

EIGHT SPEED TRANSMISSION

SHIFTER RAILS AND FORKS

All Models So Equipped

159. To remove the shifter rails and forks, first remove transmission top cover as outlined in paragraph 158 and split tractor between transmission and differential housings as outlined in paragraph 156. Unwire and remove set screws retaining shift forks and selectors to rails. Lift detent springs (Fig. 181) and plungers from housing. Remove stop plate assembly (14—Fig. 182), then withdraw shifter rails rearward and lift

Fig. 181—View showing shift rails, forks, detent springs and plungers for 8-speed transmission. Set screws (S) should be safety wired after assembly.

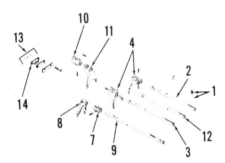

Fig. 182—Exploded view of 8-speed transmission shifter rails, forks and associated parts.

1. Detent spring & plunger
2. Second/high rail
3. Low/reverse rail
4. Shift forks
7. Selector
8. Planetary shift fork
9. Planetary shift rail
10. Fourth gear selector
11. Fourth gear shift fork
12. Fourth gear shift rail
13. Interlock balls & pin
14. Stop plate assy.

out forks and selectors. Identify shift rails and forks as they are removed to ensure correct reassembly.

When assembling rails in transmission, slide center shift rail into case and through the selector and shift fork. Tighten selector set screw and install safety wire. Rotate rail until shift fork set screw can be tightened fully, then install safety wire. Install the left shift rail and fork, then right shift rail and fork into respective bores, tighten set screws and safety wire. Slide the planetary shift rail through rear of case and locate fork over end with set screw to the outside. Slide rail forward into case and into selector. Tighten set screws and install safety wire. Be sure that interlocking pin is through the center rail and that all rails are in "NEUTRAL." Install steel balls in lock mechanism and locate at rear of case. The steel balls should be at both ends of the center rail interlock pin and the widest flat of lockplate should be over planetary rail. Tighten screws attaching the interlock mechanism to 41-47 N·m (30-35 ft.-lbs.) torque. Locate the planetary coupler around shift fork, then slide coupler into planetary unit while piloting shift fork over the rail. Tighten and safety wire the set screws.

PLANETARY UNIT

All Models So Equipped

160. R&R AND OVERHAUL. To remove the planetary unit, first separate transmission from differential housing as outlined in paragraph 156. Remove shift fork and coupler from rear of planetary. Remove the four retaining cap screws and withdraw rear cover (13—Fig. 183 or Fig. 184), rear thrust ring (6) and planet carrier (7). Pry planetary ring gear (3) with dowels from rear of housing. Remove planetary front cover (2) and shim (1).

To disassemble planet carrier (7), first remove snap ring (5) if used. Press planet pins (8) forward out of carrier. Remove planet gears (10) with needle rollers (12) and thrust washers (9).

When reassembling, be sure to account for all of the needle rollers (12). On Model MF255, there are two rows of 27 rollers each separated by three spacer washers (11—Fig. 184). On all other models, there are two rows of 16 rollers each separated by one spacer washer (11—Fig. 183). Use petroleum jelly to hold rollers in place while pressing planet pins into carrier. Gap in snap

ring (5) should be located between planet pins.

When reinstalling planetary, be sure that front plate (1—Fig. 183 or Fig. 184) and rear plate (13) are positioned with grooved side toward pinion carrier. Be sure that slots in front plate and shim (1) are aligned and are toward the top when installed on transmission. Use petroleum jelly to hold thrust rings (6) in place. Make certain that tangs of thrust rings engage notches in pinion carrier and that brass side faces away from carrier. The cutaway section or slanted section of rear plate (Fig. 185) should be positioned at lower left-hand corner of planetary, and no lockwasher

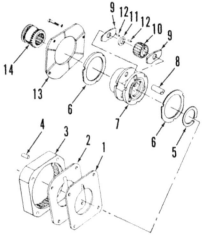

Fig. 183—Exploded view of high/low range planetary unit used on all models except MF255. See Fig. 184 for MF255 models.

1. Front shim
2. Front plate
3. Ring gear
4. Dowel
5. Snap ring
6. Thrust washer
7. Planet carrier
8. Pinion shaft
9. Side washer
10. Pinion
11. Spacer washer
12. Needle rollers
13. Rear plate
14. Shift coupler

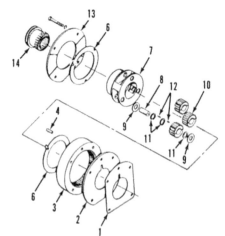

Fig. 184—Exploded view of high/low range planetary unit used on MF255 models. Refer to Fig. 183 for legend.

Fig. 185—The cutaway (slanted) section of rear plate should be over the lower left mounting screw.

Lubricate seals and bearings prior to reassembly. Seal protectors should be used when sliding input shaft into retainer and when installing pto shaft and retainer assembly over transmission input shaft. Lightly coat threads of retainer cap screws with nonhardening sealant, then tighten to 60 N·m (45 ft.-lbs.) torque. Remainder of installation is reverse of removal procedure.

TRANSMISSION INPUT SHAFT

All Models So Equipped

162. To remove transmission input shaft (11—Fig. 187), transmission must first be removed from tractor as outlined in paragraph 157. Remove shift rails and forks as outlined in paragraph 159 and remove pto input shaft as out-

is to be used on lower left-hand mounting bolt. Tighten mounting bolts evenly to a torque of 47 N·m (35 ft.-lbs.). Complete installation by reversing the removal procedure.

depth in retainer housing. When installing new bearing (5) on input shaft, be sure shielded side of bearing faces the gear teeth.

PTO INPUT SHAFT AND RETAINER HOUSING

All Models So Equipped

161. To remove the pto input shaft (8—Fig. 186) and retainer, first split tractor between engine and transmission as outlined in paragraph 148. Remove clutch release bearing, fork and shafts and the brake cross shaft (if so equipped). Disconnect hydraulic line to ipto regulating valve (if so equipped). Remove mounting cap screws from retainer and withdraw retainer with pto input shaft as a unit.

To disassemble, remove snap ring (6—Fig. 187) from rear of retainer, then push pto input shaft (8) with bearing rearward from housing. Remove oil seals (1 and 9) and needle bearing (2) if necessary.

Both oil seals should be installed with lips toward the rear. Inner seal (9) should seat against shoulder in bore of pto input shaft. Special tools (MFN850 and MFN850A) are recommended for installing the bearing and seal to correct

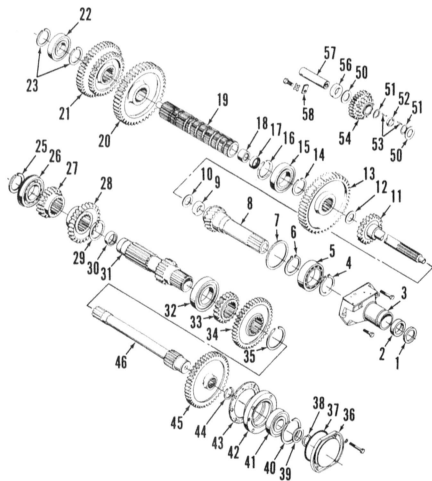

Fig. 187—Exploded view of 8-speed transmission shafts, gears and related components.

1. Oil seal	14. Snap ring	28. Third gear (33T)	42. Retainer
2. Bearing	15. Bearing	29. Snap ring	43. Gasket
3. Retainer housing	16. Snap ring	30. Bearing	44. Snap ring
4. Snap ring	17. Spacer	31. Countershaft	45. Pto drive gear
5. Bearing	18. Bearing	32. Bearing	46. Pto lower drive
6. Snap ring	19. Mainshaft	33. Fourth gear (36T)	shaft
7. "O" ring	20. Gear (1st & rev.)	34. Constant mesh gear	50. Thrust washers
8. Pto input shaft	21. Gear cluster (2nd &	35. Snap ring	51. Washers
9. Oil seal	3rd)	36. Cover plate	52. Spacer
10. Thrust washer	22. Bearing	37. "O" ring	53. Needle rollers
11. Transmission input	23. Snap rings	38. Snap ring	54. Reverse idler gear
shaft	25. Snap ring	39. Spacer	56. Spacer
12. Spacer	26. Bearing	40. Snap ring	57. Idler shaft
13. Gear (4th)	27. Second gear (23T)	41. Bearing	58. Retaining clip

Fig. 186—Front view of transmission case showing pto input shaft (8) and retainer housing (3) as a unit. Transmission input shaft (11) is removed rearward from transmission case. Pressure regulator valve (P) is not used on all models.

lined in paragraph 161. Remove cover plate (36), snap ring (38) and washer (39) from front of pto lower shaft (46). Thread two 3/8 inch UNC cap screws into tapped holes of bearing retainer (42) and tighten evenly to pull retainer with bearing from front of pto shaft. Pull pto shaft rearward to allow pto drive gear (45) to drop to bottom of housing. Remove mainshaft front sliding gear (Fig. 188), then withdraw input shaft from housing as shown in Fig. 189.

When reinstalling, make sure that grooves of rear thrust washer are toward the rear as shown in Fig. 190. Be sure to use appropriate seal protector over splines of input shaft when reinstalling pto input shaft and retainer.

MAIN (OUTPUT) SHAFT AND GEARS

All Models So Equipped

163. To remove mainshaft (31—Fig. 187), first remove transmission as outlined in paragraph 157. Remove shifter rails and forks as outlined in paragraph 159, remove planetary unit as outlined in paragraph 160 and remove main input shaft as outlined in paragraph 162. Pry or drive mainshaft rearward until front and rear bearings are free of housing bores. Withdraw sliding gear (13) from front of shaft. Remove snap ring (14), bearing (15) and snap ring (16) from front of mainshaft, then slide the shaft rearward from housing and lift out gears (20 and 21). Note that rear bearing (22) must be removed forward off mainshaft.

The inner bearing (18) in front end of shaft should be pressed into bore until flush with counterbore. The rear bearing (22) should be installed with outer snap ring toward the rear. Insert mainshaft from the rear while placing gears onto shaft. Tap shaft into forward bearing (15) until rear snap ring (16) contacts bearing, then pull shaft and bearing rearward and install front snap ring (14). Tap shaft and bearings forward into bores in housing and install sliding gear

(13) onto front of shaft. Complete installation by reversing the removal procedure.

PTO LOWER SHAFT AND DRIVE GEAR

All Models So Equipped

164. To remove the pto lower shaft (46—Fig. 187), the transmission must first be removed as outlined in paragraph 157. The shift rails and forks, pto input shaft and transmission input shaft must be removed as outlined in previous paragraphs in order to remove pto drive gear (45).

Remove brake cross shaft (if so equipped). Remove cover plate (36), snap ring (38) and washer (39) from front of shaft. Thread two 3/8 inch UNC cap screws into tapped holes in bearing retainer as shown in Fig. 191 and tighten evenly to pull retainer with bearing from front of shaft. Pull pto shaft rearward from housing and lift drive gear out top of housing.

When assembling, make sure that snap ring (44—Fig. 187) is installed in inner bore of drive gear (45) and hub of gear faces forward. Use a suitable forcing screw and crossbar as shown in Fig. 192 to pull pto shaft into front bearing. Apply a nonhardening sealer to threads of retainer cap screws, install cover

plate with new "O" ring and tighten cap screws to a torque of 60 N·m (45 ft.-lbs.).

COUNTERSHAFT

All Models So Equipped

165. To remove the transmission countershaft (31—Fig. 187) and related gears, first remove transmission as outlined in paragraph 157. Remove mainshaft as outlined in paragraph 163 and remove pto lower shaft as outlined in paragraph 164. Remove snap rings (25 and 35) from rear and front of countershaft, then bump the shaft forward until free from rear bearing. Unseat snap ring (29) from groove in center of shaft and move it forward onto unsplined section of shaft. Slide shaft rearward and remove gears (34 and 33) from front of shaft. Slide shaft forward through front of housing and remove rear gears (27 and 28).

Snap ring (29) should be around shaft in unsplined area and front bearing should be in place on shaft before installing shaft in housing. Insert countershaft through front of housing and position gears on shaft. Slide center snap ring (29) into its groove. Block front of shaft, then tap rear bearing onto

Fig. 191—Thread two 3/8 inch UNC cap screws into tapped holes in bearing retainer and tighten evenly to pull retainer with bearing from pto lower shaft.

Fig. 189—Transmission input shaft is withdrawn rearward through top opening as shown.

Fig. 188—Groove for shift fork should be toward the rear on mainshaft front sliding gear.

Fig. 190—Grooved side of input shaft thrust washer should face rearward.

Fig. 192—Pto shaft is drilled and tapped at front end for installation using a suitable puller screw and crossbar as shown.

shaft as shown in Fig. 193 until rear snap ring can be installed. Install snap ring at front of shaft. Complete installation by reversing the removal procedure.

REVERSE IDLER ASSEMBLY

All Models So Equipped

166. The reverse idler shaft (57—Fig. 187) and gear (54) can be removed from

transmission housing after removing the transmission mainshaft as outlined in paragraph 163.

The idler shaft is retained in housing bore by a clip (58) and cap screw. The idler gear contains two rows of loose needle rollers (28 rollers each row) which use the shaft and gear as inner and outer races. Removal and installation is facilitated by using a dummy shaft (1—Fig. 194) to retain rollers in

gear. The dummy shaft should be 55 mm (2-3/16 inches) long by 25 mm (1 inch) diameter. Push the dummy shaft into front of idler gear, displacing idler shaft toward the rear. Lift out gear, washers and spacer along with dummy shaft and needle rollers as a unit.

To reinstall, reverse the removal procedure.

Fig. 193—Block front of countershaft as shown when installing rear bearing onto shaft. Use a suitable driver that will contact inner race of bearing only.

Fig. 194—Use a dummy shaft (1) to hold needle rollers, washers (2), spacer (3) and idler gear (5) in place as idler shaft (4) is removed and installed.

MULTI-POWER TRANSMISSION

SHIFTERS RAILS AND FORKS

All Models So Equipped

167. To remove the shifter rails and forks, first remove transmission top cover as outlined in paragraph 158 and split tractor between transmission and differential housings as outlined in paragraph 156. Unwire and remove the set screws retaining selector and shifter forks to rails. Remove detent springs and plungers (1—Fig. 195) from front of rails. Remove shift rail stop plate and in-

terlock assembly (Fig. 196) from rear of housing, then withdraw shifter rails and forks from transmission housing. Identify shift rails and forks as they are removed to ensure correct reassembly.

When installing shift rails (2 and 3—Fig. 195) be sure that milled flat surfaces are to the rear and facing up and that interlock grooves in rear of rails face each other. Place shift rails in neutral and install interlock and stop plate assembly on rear of housing. Complete installation making sure all set screws are safety wired to prevent loosening.

PLANETARY UNIT

All Models So Equipped

168. R&R AND OVERHAUL. To remove the planetary unit, first separate transmission from differential housing as outlined in paragraph 156. Remove shift fork (Fig. 199) and coupler from rear of planetary. Remove the four retaining cap screws and withdraw rear cover (13—Fig. 197 or Fig. 198), rear thrust ring (6) and planet carrier (7). Pry planetary ring gear (3) with dowels from rear of housing. Remove front plate (2) and shim (1).

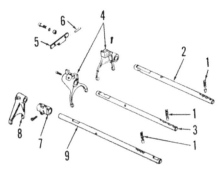

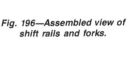

Fig. 196—Assembled view of shift rails and forks.

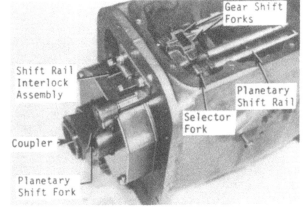

Fig. 195—Exploded view of shift rails, forks and associated parts typical of Multi-Power models.

1. Shift detents
2. Second/third rail
3. First/reverse rail
4. Shift forks
5. Stop plate
6. Interlock pin
7. Selector
8. Planetary shift fork
9. Planetary shift rail

To disassemble planet carrier, first remove snap ring (5), if used, from carrier. Press planet pins (8) forward out of carrier. Remove planet gears (10) with needle rollers (12) and thrust washers.

When reassembling, be sure to account for all of the needle rollers (12). On Models MF255, there are two rows of 27 rollers each and three spacer washers (11—Fig. 198). On all other models, there are two rows of 16 rollers each separated by one spacer washer (11—Fig. 197). Use petroleum jelly to hold rollers in place while pressing planet pins into carrier. Install snap ring (5), if equipped, so gap in ring is located between two of the pins.

When reinstalling planetary, be sure that front plate (2—Fig. 197 or Fig. 198) and rear plate (13) are positioned with grooved side facing planet carrier. Be sure that slots in front plate and shim (1) are aligned and are toward the top when installed on transmission housing. Use petroleum jelly to hold thrust rings (6) in place. Make certain that tangs of thrust rings engage notches in planet carrier and that brass side of rings face away from carrier. The cutaway section or slanted section of rear plate (Fig. 199) should be positioned at lower left-hand corner of planetary, and no lockwasher is to be used on lower left-hand mounting bolt. Tighten mounting bolts evenly to a torque of 47 N·m (35 ft.- lbs.). Complete installation by reversing the removal procedure.

PTO INPUT SHAFT AND RETAINING HOUSING

All Models So Equipped

169. To remove pto input shaft and retainer (Fig. 200), first separate engine from transmission housing as outlined in paragraph 148. Remove clutch release bearing, release fork and pivot shafts. Disconnect shift linkage and remove cap screw retaining shift lever bracket. Disconnect hydraulic line from shift control valve. Remove cap screws attaching pto input shaft retainer to transmission housing, then withdraw retainer, pto input shaft and shift valve as a unit.

Unbolt and remove shift control valve from retainer housing. Remove large snap ring (29—Fig. 203) from rear of retainer, then bump pto input shaft (32) with bearing rearward from retainer housing. Remove oil seals (30 and 38), needle bearing (37), seal rings (31) and ball bearing (33) if necessary.

Both oil seals (30 and 38) should be installed with lips facing rearward. Special tools (MFN850 and MFN850A) are available for installing needle bearing (37) and front seal to correct depth in retainer. Be sure shielded side of ball bearing (33) is toward input shaft gear teeth. Install new seal rings (31) and lubricate all parts with clean transmission oil.

Assemble shift control valve onto retainer housing and tighten mounting screws evenly. Suitable seal protectors should be used over shaft splines when sliding pto input shaft into retainer and when installing pto shaft and retainer assembly over main input shaft. Lightly coat threads of retainer housing cap screws with nonhardening sealant, then tighten cap screws to 60 N·m (45 ft.-lbs.) torque. Remainder of assembly is reverse of disassembly procedure.

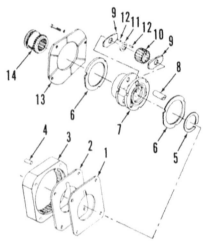

Fig. 197—Exploded view of high/low range planetary unit used on all models except MF255. See Fig. 198 for MF255 models.

1. Front shim	8. Pinion shaft
2. Front plate	9. Side washer
3. Ring gear	10. Pinion
4. Dowel	11. Spacer washer
5. Snap ring	12. Needle rollers
6. Thrust washer	13. Rear plate
7. Planet carrier	14. Shift coupler

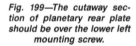

Fig. 199—The cutaway section of planetary rear plate should be over the lower left mounting screw.

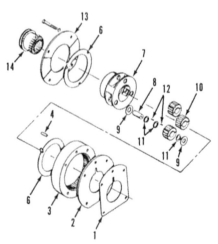

Fig. 198—Exploded view of high/low range planetary unit used on MF255 models. Refer to Fig. 197 for legend.

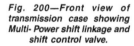

Fig. 200—Front view of transmission case showing Multi- Power shift linkage and shift control valve.

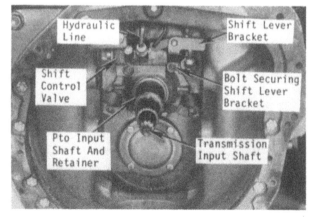

SHIFT CONTROL VALVE

All Models So Equipped

170. The Multi-Power shift control valve is removed with the pto input shaft and retainer as outlined in paragraph 169. To disassemble the removed valve, remove inlet fitting (10—Fig. 201) and detent assembly (12). Withdraw shift valve (8) from valve body. Remove plug (1), spring (3) and regulator valve (4). Remove spring (5) and ball (6) if used.

Examine all parts for wear, scoring or other damage and renew if necessary. All parts are available individually. Renew all "O" rings and gasket.

When reassembling valve, make sure that regulating valve spool (4) is installed with "V" notched end first into valve body. Tighten plug (1) to a torque of 27 N·m (20 ft.-lbs.). Install shift valve spool (8) before fitting (10) is installed.

To reinstall, reverse the removal procedure. Use a suitable seal protector over splines of main input shaft to avoid damaging pto input shaft oil seal.

The control valve spool must move fully to high and low positions as Multi-Power control lever (10—Fig. 202) is moved to top and bottom of control slot. To adjust shift linkage, move hand lever toward "HIGH" position until lever is within 1.5 mm (1/16 inch) from top of slot. Loosen clamp bolt securing link (6) to lower rod (4), then push lower control rod (4) downward as far as it will go. Retighten clamp bolt to lock the adjustment.

Refer to paragraph 242 for Multi-Power pressure test procedure.

MAIN (OUTPUT) SHAFT

All Models So Equipped

171. To remove mainshaft (14—Fig. 203) and related gears, first separate tractor between transmission housing and differential housing as outlined in paragraph 156. Remove shifter rails and forks as outlined in paragraph 167 and

remove planetary unit as outlined in paragraph 168. Pry mainshaft rearward until front bearing (15) is clear of bore in transmission case. Remove snap ring (16) from front of mainshaft, then pull shaft rearward in a slide hammer action to bump bearing from front of shaft. Withdraw shaft rearward from housing while lifting gears (12 and 13) from housing. Note that rear bearing (11) must be removed forward off mainshaft.

When installing mainshaft, be sure that cluster gear (12) is positioned with smaller gear toward the front and that first/reverse gear (13) is positioned with shift fork hub facing rearward. Use special tool (MFN458) or similar thin

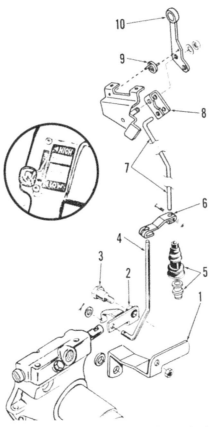

Fig. 202—Exploded view of Multi-Power clutch control linkage.

1. Bracket	6. Link
2. Shift lever	7. Upper link rod
3. Pivot bolt	8. Spacer
4. Lower link rod	9. Spring
5. Bushing & boot	10. Lever

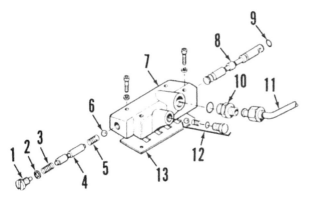

Fig. 201—Exploded view of Multi-Power shift control valve. Spring (5) and ball (6) are not used on some models.

1. Plug
2. Gasket
3. Spring
4. Regulating valve
5. Spring
6. Ball
7. Valve body
8. Shift valve
9. "O" ring
10. Fitting
11. Hydraulic line
12. Detent assy.
13. Gasket

spacer plate to block front bearing (Fig. 205) as mainshaft is tapped into the bearing. Remove spacer plate, install front snap ring and bump shaft and bearings forward into position. Complete installation by reversing removal procedure.

TRANSMISSION INPUT SHAFT AND MULTI-POWER CLUTCH

All Models So Equipped

172. REMOVE AND REINSTALL. To remove transmission input shaft (23—Fig. 203) and Multi-Power clutch unit (27), first remove transmission as outlined in paragraph 157. Remove shifter rails and forks, planetary unit, pto input shaft and retainer and mainshaft as outlined in previous paragraphs. Remove seal rings (22) and thrust washer (24) from front of input shaft. Remove spacer (17) and main drive pinion gear (19) from rear of shaft. Withdraw input shaft rearward while lifting clutch (27) and associated parts out top opening.

Examine all parts for wear or scoring and renew if necessary. Overhaul the removed Multi-Power clutch unit as outlined in paragraph 173.

To reinstall input shaft and clutch unit, reverse the removal procedure. Be sure that tab of spacer (17) is correctly located and that convex side of spacer is toward the clutch unit. Make certain that bronze side of thrust washer (Fig. 206) faces away from Multi-Power clutch pack and that ends of input shaft seal rings are properly hooked. A suitable seal protector should be used over end of input shaft to prevent damage to oil seal when assembling.

173. MULTI-POWER CLUTCH. To disassemble the removed clutch unit, place unit on a bench and lift out overdrive pinion (25—Fig. 203). Push clutch retainer plate (2—Fig. 207) inward and remove snap ring (1). Remove retainer plate, friction discs (4), plates (5) and springs (3). Remove clutch piston (6) from housing (9) and remove piston seal rings (7 and 8).

Inspect all parts for wear, distortion or other damage and renew if necessary. The friction discs and plates must be renewed as a complete set. Refer to the following specifications:

Friction Discs—
Thickness . . . . .2.41-2.59 mm
(0.095-0.102 in.)
Oil Groove Depth . . . . . .0.38-0.63 mm
(0.015-0.025 in.)

Maximum Allowable
Distortion . . . . .2.92 mm
(0.115 in.)

Clutch Plates—

Thickness . . . . 1.67-1.75 mm
(0.066-0.069 in.)

Maximum Allowable
Distortion . . . . 2.21 mm
(0.088 in.)

Release Springs—

Free Length . . . . 17.8 mm
(0.70 in.)

Test Length . . . . 12.7 mm
(0.50 in.)

Test Load . . . . 29.3-35.5 N
(6.6-8.0 lbs.)

Renew piston sealing rings (7 and 8) whenever clutch is disassembled. When installing the piston, carefully compress the outer sealing ring (8) using a narrow blade screwdriver or similar tool and work the piston into its bore. The inner ring (7) will normally compress because of chamfer in inner bore of piston if care is used in assembly.

When assembling clutch plates, note that clutch drum (9) contains six bleed holes which are evenly spaced, and that driving plates (5) have six external driving lugs. With piston installed, refer to Fig. 208 and install the first drive plate on top of piston with lugs (L) one spline clockwise from bleed holes (B). Install an internally splined clutch disc (D), then the second plate with drive lug (L) one spline clockwise from lug on first plate as shown. Repeat the procedure for remainder of discs and plates. Place the piston return springs (3—Fig. 207) on the driving lugs of first plate installed, then install retainer plate (2) and snap ring (1). Assemble input overdrive pinion into clutch assembly.

PTO LOWER SHAFT AND DRIVE GEAR

All Models So Equipped

174. To remove the pto lower shaft (55—Fig. 203) and drive gear (56), first remove transmission assembly as outlined in paragraph 157. Remove clutch release bearing, fork and shafts and the brake cross shaft if so equipped. Remove mainshaft and gears, pto input shaft and transmission input shaft as outlined in previous paragraph.

Remove cover plate (65—Fig. 203) from front of housing and remove snap ring (63) and thrust washer (62) from front of pto shaft. Thread two 3/8 inch UNC cap screws into tapped holes in retainer housing (59) and tighten evenly to pull retainer with bearing from pto shaft. Remove pto shaft rearward from housing and lift drive gear out top opening of housing.

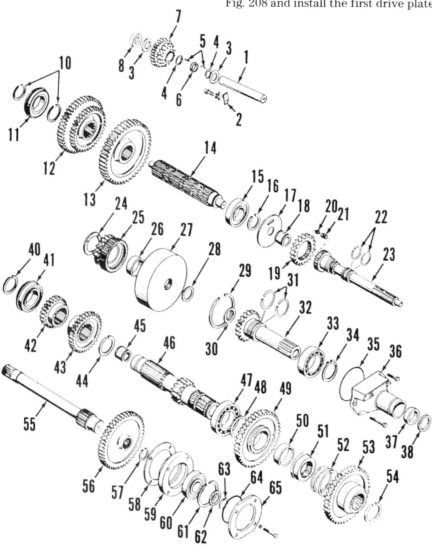

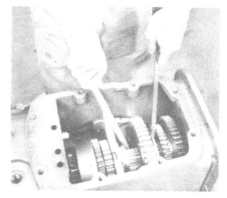

Fig. 204—Pry mainshaft rearward as shown to free front bearing from bore in transmission case.

Fig. 205—A special tool (1), MFN458, or similar thin spacer plate is required for installing mainshaft (2) into front bearing (3).

Fig. 203—Exploded view of typical shaft, gears and related components used in Multi-Power transmission.

1. Idler shaft
2. Retainer clip
3. Thrust washers
4. Washers
5. Needle rollers
6. Spacer
7. Reverse idler gear
8. Spacer
10. Snap rings
11. Bearing
12. Cluster gear (2nd & 3rd)
13. Gear (1st & rev.)
14. Mainshaft
15. Bearing
16. Snap ring
17. Spacer
18. Bearing
19. Direct drive pinion
20. Thrust plug
21. Spring
22. Seal rings
23. Transmission input shaft
24. Thrust washer
25. Overdrive pinion
26. Bushing
27. Multi-Power clutch assy.
28. Thrust washer
29. Snap ring
30. Oil seal
31. Seal rings
32. Pto input shaft
33. Bearing
34. Snap ring
35. "O" ring
36. Retainer housing
37. Bearing
38. Oil seal
40. Snap ring
41. Bearing
42. Gear
43. Gear
44. Snap ring
45. Bearing
46. Countershaft
47. Bearing
48. Thrust washer
49. Direct drive gear
50. Bushing
51. Jaw coupler
52. Spring
53. Overdrive gear
54. Snap ring
55. Pto lower shaft
56. Pto drive gear
57. Snap ring
58. Gasket
59. Bearing retainer
60. Bearing
61. Snap ring
62. Washer
63. Snap ring
64. "O" ring
65. Cover plate

Fig. 206—Bronze side of input shaft thrust washer should face away from clutch pack. Use appropriate seal protector over end of shaft when assembling.

When assembling, make sure that snap ring (57) is installed in inner bore of drive gear and that hub side of gear faces forward. Front end of pto shaft contains a threaded hole which can be used to pull shaft into front bearing as shown in Fig. 209. Apply nonhardening sealer to threads of retainer cap screws, install cover plate with new "O" ring and tighten cap screws evenly to 60 N·m (45 ft.- lbs.) torque.

COUNTERSHAFT

All Models So Equipped

175. To remove transmission countershaft (46—Fig. 203) and related gears,

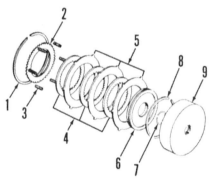

Fig. 207—Exploded view of Multi-Power clutch assembly.

1. Snap ring
2. Retaining plate
3. Release springs
4. Clutch discs
5. Drive plates
6. Piston
7. Seal ring
8. Seal ring
9. Clutch housing

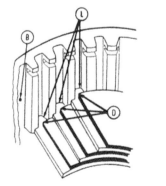

Fig. 208—Cross-sectional view of Multi-Power clutch showing recommended method of assembling clutch discs (D) and clutch plate drive lugs (L) in relation to bleed holes (B) in clutch housing.

first remove transmission as outlined in paragraph 157. Remove mainshaft and gears, pto input shaft, transmission input shaft and pto lower shaft as outlined in previous paragraphs.

Remove snap ring (40) from rear of countershaft, then bump the shaft forward until free from rear bearing. Unseat snap ring (44—Fig. 210) from groove in center of shaft and move it forward onto unsplined section of shaft. Remove snap ring (54—Fig. 203) from front of shaft. Tap shaft rearward and remove overdrive gear (53), spring (52), jaw coupler (51), direct drive gear (49) and thrust washer (48) out top opening of housing. Bump shaft forward out of housing and lift out gears (42 and 43).

Snap ring (44—Fig. 210) should be around shaft in unsplined area and front bearing should be in place on countershaft before installing into housing. Insert countershaft from the front and position the two gears (42 and 43) on rear of shaft with their hubs facing together. Move shaft rearward to allow installation of thrust washer, direct drive gear, coupler, spring and overdrive gear onto front of shaft. Move shaft forward and push second and third gears rearward to allow center snap ring to be seated in its groove. Block front of shaft, then tap rear bearing into housing bore and onto shaft until rear snap ring can

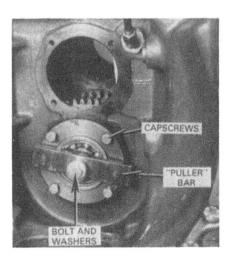

Fig. 209—Pto lower shaft is drilled and tapped at front end for installation as shown.

be installed. Complete installation by reversing the removal procedure.

REVERSE IDLER ASSEMBLY

All Models So Equipped

176. The reverse idler shaft (1—Fig. 203) and gear (7) can be removed from transmission housing after removing the transmission mainshaft as outlined in paragraph 171. The idler shaft is retained in housing by a clip (2) and cap screw at rear of housing.

The idler gear contains two rows of loose needle rollers (28 rollers each row) which use the shaft and gear as inner and outer races. Removal and installation is facilitated by using a dummy shaft (Fig. 211) to retain rollers in gear. The dummy shaft should be 55 mm (2-3/16 inches) long by 25 mm (1 inch) diameter. Push the dummy shaft into front of idler gear, displacing idler shaft toward the rear. Lift out gear, washers and spacer along with dummy shaft and needle rollers as a unit.

To reinstall, reverse the removal procedure.

Fig. 210—View of countershaft and gears partially assembled. Be sure hubs of gears (42 and 43) face each other.

Fig. 211—View of reverse idler gear and related parts. Use a dummy shaft to hold needle rollers in place in gear when removing and installing.

DIFFERENTIAL AND BEVEL DRIVE GEARS

DIFFERENTIAL

All Models

177. R&R AND OVERHAUL. The differential assembly can be removed after removing the complete final drive unit from left-hand side of tractor as outlined in paragraph 188.

To disassemble the removed unit, remove bearing cone (7—Fig. 215) from right-hand side of differential. Remove the eight through-bolts and lift off coupler cap (8), shims (if used) and differential case half (9). Remove axle gears (11) with thrust washers (10) and cross shaft (12) with pinions (13) and thrust washers (14).

Examine all parts and renew any that are excessively worn or damaged. Renew axle gears (11) in pairs and differential pinions (13) in sets of four. Backlash of differential gears is controlled by thrust washers (10 and 14). Refer to paragraphs 182 and 183 for renewal of bevel ring gear and pinion (15).

When reassembling differential unit, make sure that identification markings on case halves (9 and 16) are aligned.

Tighten through-bolts to a torque of 108 N·m (80 ft.-lbs.).

Differential carrier bearing preload will need to be checked when differential lock coupler cap (8), differential case halves (9 and 16) and/or carrier bearings are renewed. Refer to paragraph 178 for adjustment procedure.

To reinstall differential assembly, reverse the removal procedure.

178. BEARING PRELOAD. Differential carrier bearing preload will need to be checked when differential case halves (9 and 16—Fig. 215), coupler cap (8) and/or carrier bearings are renewed. To check adjustment, differential assembly and left axle housing should be installed, and right axle housing and hydraulic lift cover removed. Preload is adjusted by installing a selective thickness spacer shield (3) between bearing cup (4) and right-hand carrier plate (1). On some models, shims are also available which may be installed between coupling cap (8) and right-hand differential case half (9) to adjust preload.

NOTE: When adjusting bearing preload using the following procedure, two spacer shield changes are likely. Shield is normally distorted beyond reuse when it is removed.

Unbolt and remove carrier plate (1) from right axle housing. Install bearing cup (4) and a 1.25 mm (0.050 inch) thick spacer shield in carrier plate. Install carrier plate (without "O" ring) on differential housing using three equally spaced cap screws. Tighten cap screws evenly and carefully until carrier bearing end play is just removed. Differential should be rotated to center the bearings and end play checked through lift cover opening in top of differential housing as cap screws are being tightened.

Measure the clearance between carrier plate and differential housing at several locations and average the clearances. The thickness of installed spacer shield (and preload) is correct when average clearance is within range of 0.25-0.38 mm (0.010-0.015 inch). If clearance is excessive, remove bearing cap and shield and install new shield of appropriate smaller thickness. If clearance is less than 0.25 mm (0.010 inch), check for improper assembly. If unit is correctly assembled, install a thicker spacer shield or a shim (if available) between coupler cap (8) and case half (9). Be sure spacer shield is installed with concave side (Fig. 216) toward the differential.

DIFFERENTIAL LOCK

All Models

180. ADJUSTMENT. The differential lock pedal arm (11—Fig. 220) should be against stop (S) when differential lock is disengaged. To adjust, loosen locknut (8) and turn clevis (10) until pedal arm

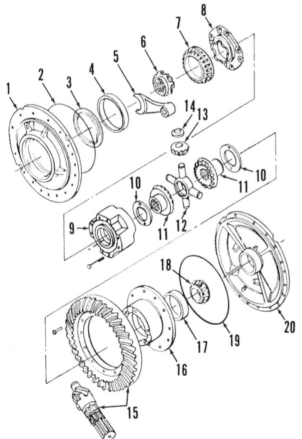

Fig. 215—Exploded view of differential assembly typical of all models. Spacer shield (3) is available in different thicknesses to adjust carrier bearing preload.

1. Carrier plate, R.H.
2. "O" ring
3. Spacer shield
4. Bearing cup
5. Shift fork
6. Differential lock coupler
7. Bearing cone
8. Coupler cap
9. Differential case half
10. Thrust washers
11. Axle gears
12. Pinion cross
13. Pinion gear
14. Thrust washer
15. Bevel ring gear & pinion
16. Differential case half
17. Bearing cup
18. Bearing cone
19. "O" ring
20. Carrier plate, L.H.

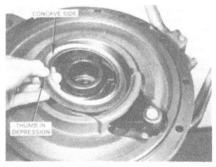

Fig. 216—Spacer shield is correctly installed with concave (dished) side facing the differential.

just contacts the stop. Tighten locknut to secure adjustment.

181. REMOVE AND REINSTALL. To remove differential lock actuating assembly, first remove right-hand final drive unit as outlined in paragraph 188. Drive spring pin out of shift fork (2—Fig. 219), remove two screws attaching carrier plate to axle housing, then remove carrier plate, fork and jaw coupler (1) from axle housing. Remove spring (3), washer (4) and snap ring (5) from actuating shaft (6). Loosen locknut (8) and unscrew shaft from clevis (10). To remove inner coupler half (8—Fig. 215), differential carrier bearing cone (7) must first be removed using a suitable puller. Note that cap screws retaining the inner coupler half also hold the differential case halves together.

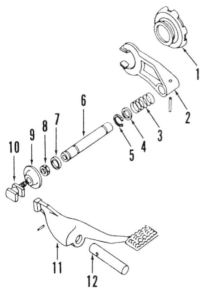

Fig. 219—Exploded view of differential lock actuating mechanism used on models with dry brakes. Models with wet brakes are similar except that an "O" ring is used in place of oil seal (7).

1. Jaw coupler	7. Oil seal
2. Shift fork	8. Locknut
3. Spring	9. Boot
4. Washer	10. Clevis
5. Snap ring	11. Pedal arm
6. Actuating shaft	12. Pivot pin

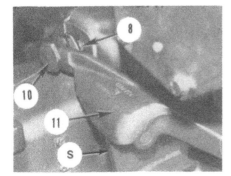

Fig. 220—Differential lock pedal arm (11) should just contact stop (S) when differential lock is disengaged. Refer to text for adjustment.

To reinstall, reverse the removal procedure while noting the following special instructions: Install new actuating shaft oil seal (7—Fig. 219) and axle shaft oil seal (models so equipped) into carrier plate with seal lips facing toward differential. Be sure that recess in washer (4) fits over snap ring (5). Tighten inner coupler half retaining cap screws to a torque of 108 N·m (80 ft.-lbs.). If inner coupler half and/or differential carrier bearing was renewed, adjust carrier bearing preload as outlined in paragraph 178. Adjust differential lock as outlined in paragraph 180.

BEVEL DRIVE GEARS

All Models

182. BEVEL PINION. The main drive bevel pinion is available only as a matched set with the bevel ring gear. Refer to paragraph 183 for renewal of bevel ring gear.

To remove bevel pinion on models equipped with Multi-Power, independent pto and/or auxiliary hydraulic system, it will be necessary to split the tractor between transmission housing and differential housing as outlined in paragraph 156. Remove hydraulic lift cover as outlined in paragraph 215. Remove hydraulic pumps as outlined in paragraph 222 and withdraw ipto clutch (if so equipped) as shown in Fig. 221. If equipped with two-speed ipto, remove snap ring (16—Fig. 222) and withdraw reduction gear cluster (19) with needle bearings and thrust washer from front of pinion shaft (4). On all models, remove cap screws attaching bearing retainer (8—Fig. 223 or 224) to differential housing, then thread two cap screws into tapped holes of retainer and tighten evenly to force retainer with bevel pinion assembly from housing.

On models equipped with live pto (without auxiliary hydraulic pump), the hydraulic lift pump and bevel pinion can be removed through top opening in differential housing after hydraulic lift cover is removed. However, some

Fig. 221—On models equipped with independent pto, remove hydraulic pumps out front opening of housing and withdraw ipto clutch unit as shown.

mechanics prefer to split the tractor and remove hydraulic pump through front opening of differential housing. The following procedure describes removal through top opening in differential housing.

Remove hydraulic lift cover as outlined in paragraph 215. Remove pto shift cover from left side of housing. Collapse and remove pinion drive shaft and coupler. Remove the pto output shaft as outlined in paragraph 200. Remove hydraulic pump retaining pins from each side of the differential housing, then lift pump out top opening of housing. Remove snap ring (16—Fig. 223 or 224) and pto ground speed drive gear (15) from pinion shaft. Remove six cap screws retaining bearing carrier (8) in differential housing, then thread two cap screws into tapped holes in carrier to force carrier with bevel pinion from housing bore.

To disassemble the removed bevel pinion unit, proceed as follows: On early models using a lockwasher (14—Fig. 224) to retain adjusting nut (13), bend back locking tabs and remove nut and drive hub (11). On later models using two needle rollers (14—Fig. 223) to retain adjusting nut (13), it will be necessary to split the nut to remove it from pinion shaft. Be careful not to damage threads of pinion when removing the nut. Remove sleeve (12) and splined hub (11) on models equipped with single speed pto, or the bearing sleeve (21), keeper (20) and thrust washer (22) on dual speed pto models. Remove bearing carrier (8) with front bearing. Remove rear bearing cone (5) and pilot bearing (3) from pinion shaft using a suitable puller.

Inspect all parts for wear, scoring or pitting and renew as necessary. The bearing carrier, bearing cups and bearing cones are serviced as an assembly.

Fig. 222—Remove snap ring (16) and gear cluster (19) from front of pinion shaft (4). Unit shown is equipped with dual speed ipto.

If bevel pinion is renewed, bevel ring gear must also be renewed.

When assembling bevel pinion, install pilot bearing with chamfered inside diameter toward pinion gear. Be sure that inner bearing cone is seated against shoulder of pinion gear. Tighten bearing adjusting nut to obtain a preload reading of 2.1-2.4 N·m (18-22 in.-lbs.) rolling torque on pinion shaft measured as shown in Fig. 225. Tap ends of pinion shaft with a soft face hammer to center the bearings, then recheck rolling torque and readjust if necessary. Note that preload may also be checked using a spring scale with a cord wrapped around outer diameter of bearing retainer. A pull of 2.5-2.9 kg (5.5-6.5 pounds) to turn flange is equal to 2.1-2.4

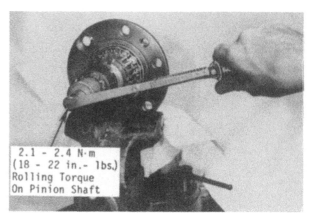

Fig. 225—Use a torque wrench to check pinion shaft rolling torque when adjusting bearing preload.

N·m (18-22 in.-lbs.) rolling torque. When preload is correct, bend tabs of lockwasher (if so equipped) into slots of

nut, or drive the needle rollers (14—Fig. 226) down each side of one of the pinion splines to lock the nut in place. Be sure that needle rollers are driven flush with end of adjusting nut.

To reinstall pinion assembly, reverse the removal procedure. Make sure that dowel pin (Fig. 227) in retainer is aligned with drilled hole in housing before pressing retainer into housing. Tighten retainer cap screws to a torque of 108 N·m (80 ft.-lbs.).

183. BEVEL RING GEAR. The main drive bevel ring gear is available only in a matched set with the bevel pinion.

To remove bevel ring gear, first remove differential assembly and disassemble differential case halves as outlined in paragraph 177. Rivets are used to attach original production ring gear to differential case half, while nuts and bolts plus

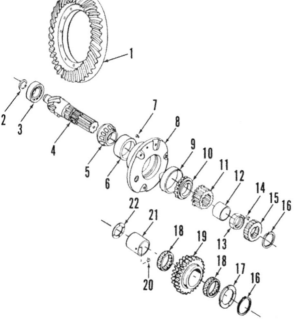

Fig. 223—Exploded view of bevel pinion and related components used on late production tractors. Items 17 through 22 are used on models equipped with dual speed independent pto.

1. Ring gear
2. Snap ring
3. Pilot bearing
4. Bevel pinion
5. Bearing cone
6. Bearing cup
7. Dowel pin
8. Bearing retainer
9. Bearing cup
10. Bearing cone
11. Drive gear hub
12. Spacer
13. Adjusting nut
14. Needle rollers
15. Live pto ground speed drive gear
16. Snap ring
17. Thrust washer
18. Needle bearings
19. Cluster gear
20. Retainer pin
21. Bearing sleeve
22. Thrust washer

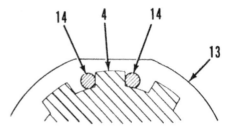

Fig. 226—On some models, bearing adjusting nut (13) is secured by driving needle rollers (14) down each side of one of the pinion splines (4). The nut must be split for removal.

Fig. 224—Exploded view of bevel pinion and related components used on early production tractors. Refer to Fig. 223 for legend except for tab lockwasher (14).

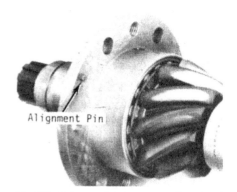

Fig. 227—Make sure that dowel pin is aligned with drilled hole in differential housing before installing retainer in housing.

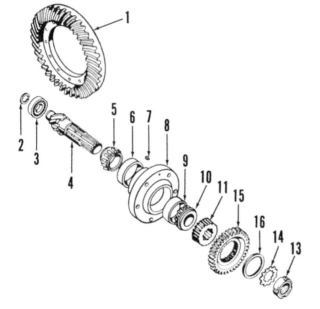

epoxy adhesive are used to retain replacement ring gear. Drill and punch out rivets (if so equipped) or remove nuts and bolts, then drive ring gear from case half.

An epoxy resin kit is available from Massey-Ferguson and should be used to bond ring gear to case half. Thoroughly degrease mating surfaces of ring gear and differential case half. Follow the instructions given in epoxy kit and apply an even coating of adhesive to mating surfaces of ring gear and case half. Note that these two components are an interference fit, therefore their bolt holes must be aligned accurately before they are assembled. Install bolts with their heads against ring gear. Apply Loctite 262 to threads, then tighten nuts to a torque of 160 N·m (120 ft.-lbs.). The epoxy must cure for a minimum of 24 hours at a temperature of 20° C (68° F) before reinstalling ring gear and differential assembly. Cure time can be reduced by increasing temperature; follow instructions given in epoxy kit.

REAR AXLE AND FINAL DRIVE

WHEEL AXLE AND PLANETARY UNIT

All Models

185. REMOVE AND REINSTALL. To remove the wheel axle and planet carrier as a unit, support rear of tractor and drain oil from final drive planetary unit. Remove wheel and tire unit and fender assembly. Apply parking brake to ensure that brake plates will remain aligned in case axle shaft is withdrawn with planetary unit. Scribe a line across drive cover (4—Fig. 230), ring gear (15) and axle housing (25) to ensure proper alignment when reassembling. Remove top bolt from drive cover and install a suitable lifting eye as shown in Fig. 228.

Remove remainder of retaining bolts and separate unit from axle housing.

To reinstall, reverse the removal procedure while making sure that previously made scribe marks are aligned. Tighten retaining nuts and bolts to a

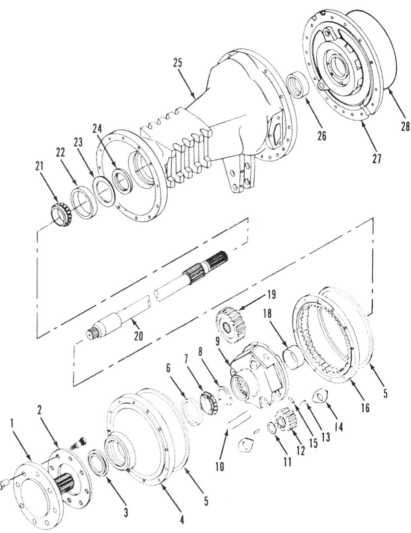

Fig. 230—Exploded view of planetary final drive and axle housing assembly used on early production models. On tractors with wet type disc brakes, oil seals (24 and 26) should not be used. Bushing (18) is not used on some models.

Fig. 228—Attach a lifting eye and hoist as shown to support wheel axle and planet carrier during removal and installation.

Fig. 229—Support axle housing as shown when removing and installing complete final drive unit.

1. Wheel axle	8. Split ring	15. Roll pin	22. Bearing cup
2. Dust shield	9. Planet carrier	16. Ring gear	23. Shim
3. Oil seal	10. Shaft	18. Bushing	24. Oil seal
4. Drive cover	11. Spacer	19. Sun gear	25. Axle housing
5. Gasket	12. Planet gear	20. Axle shaft	26. Oil seal
6. Bearing cup	13. Needle rollers	21. Bearing cone	27. Carrier plate
7. Bearing cone	14. Thrust washers		28. "O" ring

torque of 75 N·m (55 ft.-lbs.). Refill unit to proper level with Permatran 111 Oil or equivalent.

186. OVERHAUL PLANETARY UNIT. To disassemble the removed final drive unit, first remove ring gear (16—Fig. 230) from planet carrier. Drive out roll pins (15) retaining planet gear shafts (10). Thread a 3/8 inch cap screw into end of planet gear shafts and pull shafts from planet carrier. Remove planet gears (12) and thrust washers (14) being careful not to lose the needle rollers (13). Remove sun gear (19) through large opening in side of carrier. Use a suitable puller to remove bearing cone (21) from carrier hub.

Using Special Nut (MFN265) and Forcing Screw (MFN764-3) or other suitable puller, remove planet carrier from wheel axle splines as shown in Fig. 231. Remove the two half rings (8—Fig. 230), then press wheel axle (1) from drive cover (4). Remove bearing cup (6) and outer oil seal (3) from drive cover.

Coat outside diameter of new outer oil seal (3) with a suitable sealant. Press seal (with lip facing inward) into drive cover (4) until outer face of seal is 2 mm (0.080 inch) above surface of cover. Press bearing cup (6) into drive cover until it bottoms in cover counterbore. Lubricate lip of seal, then assemble wheel axle and bearing cone into drive cover. Press bearing cone onto axle until rollers are fully seated in bearing cup. Install a pair of half ring segments (8) of appropriate thickness to give a clearance of 0.00-0.025 mm (0.000-0.001 inch) between bearing cone and half rings.

Assemble sun gear and planet gears into carrier. Be sure to account for all the needle rollers (13). Each planet gear contains two rows of loose needle rollers (29 rollers in each row) separated by a washer (11). Note that large thrust washers (14) are used in large openings in carrier and small washers are used in

small openings. Use new roll pins (15) to retain shafts (10), and drive the pins in until flush with outer edge of carrier. Press new bearing cone (21), if removed, onto planet carrier hub.

To renew inner bearing cup (22), first lock the parking brake to hold brake pistons in place. Withdraw axle shaft (20), then pry bearing cup and oil seal (models with dry brakes) from axle housing. Install new seal (if used) with drain hole downward and seal lip facing outward (toward planetary). Add 0.76 mm (0.030 inch) thickness of shims to original shim pack, then install shims and bearing cup in housing.

Install the assembled planetary final drive unit omitting main axle shaft (20) and gaskets (5). Check and adjust bearing preload as outlined in paragraph 187.

When preload is properly adjusted, install main axle shaft and final drive unit using new gaskets. Be sure that stepped side of ring gear faces outward as shown in Fig. 232.

187. BEARING PRELOAD ADJUSTMENT. The planet carrier bearing preload is adjusted by means of shims installed between inner bearing cup and shoulder of axle housing (Fig. 232). Shims should be selected to provide 0.15-0.40 mm (0.006-0.016 inch) preload as follows:

With final drive unit removed, first lock the brakes to hold brake discs in place and withdraw main axle shaft (20—Fig. 230). Remove bearing cup (22) and add 0.76 mm (0.030 inch) thickness of shims to existing shim pack (23). Reinstall bearing cup and final drive unit omitting main axle shaft (20) and gasket (5). Install three equally spaced mounting bolts and tighten evenly until wheel axle end play is just removed. Bump outer end of wheel axle with a hammer to make sure that bearings are seated. Measure the gap between planetary ring gear and axle housing next to each of the three bolts and average the measurements. Remove the final drive unit and inner bearing cup.

Remove shims equal to average measured gap plus 0.25 mm (0.010 inch) to establish the recommended preload. Reinstall bearing cup, main axle shaft and final drive unit with new gaskets.

FINAL DRIVE AND AXLE HOUSING ASSEMBLY

All Models

188. REMOVE AND REINSTALL. To remove planetary final drive and axle housing as a unit, first drain oil from planetary housing and from transmission and differential housings. Suitably support rear of tractor. If both axle housings are to be removed, drive wooden wedges between front axle and axle support to prevent tipping. Remove rear wheel and tire, fender and ROPS. Disconnect brake linkage rods on models with mechanical brakes, or disconnect hydraulic brake pipes and parking brake linkage on models with hydraulic brakes. Disconnect hitch lift link at the lift arm yoke and disconnect lower lift arm at the axle housing. If right-hand axle is being removed disconnect differential lock pedal. Support axle housing assembly with a suitable hoist or floor jack, remove retaining bolts and withdraw axle housing as shown in Fig. 229. When removing left-hand axle housing, be careful that differential assembly does not fall out of center housing.

To reinstall, reverse the removal procedure while noting the following special instructions: Tighten ROPS frame mounting bolts to 170 N·m (125 ft.-lbs.) torque. Tighten wheel stud nuts evenly to a torque of 270 N·m (200 ft.-lbs.) on tractors equipped with 16.0 mm (5/8 inch) diameter studs, or to a torque of 325 N·m (240 ft.-lbs.) on tractors equipped with 17.5 mm (11/16 inch) diameter studs. Bleed hydraulic brakes as outlined in paragraph 192.

189. AXLE SHAFT AND HOUSING. Axle shaft (20—Fig. 230) can be withdrawn after removing wheel axle and planetary unit as outlined in paragraph 185. Lock the wheel brake prior to removing axle shaft to keep brake discs properly aligned.

On models equipped with dry brakes, the axle housing outer oil seal (24) can be renewed after removing axle shaft and bearing cup (22). Outer diameter of new seal should be coated with a suitable sealer and should be installed with lip facing outward (toward planetary unit).

Main axle inner oil seal (26) is located in differential carrier plate on models with dry brakes. Seal can be renewed

Fig. 231—Use special nut and forcing screw to remove planet carrier from wheel axle shaft as shown.

Fig. 232—Planetary ring gear must be positioned with stepped side outward.

after removing axle housing from differential housing. Coat outer diameter of new seal with sealant before installing. Make sure that seal lip faces away from brake contact surface of carrier (Fig. 233) and oil drain holes are aligned.

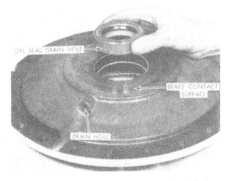

Fig. 233—On models with dry brakes, drain holes in oil seal and carrier plate must be aligned when seal is installed. Both seal lips face away from brake contact surfaces.

BRAKES

All models are equipped with disc type brakes. Brake assemblies on some tractors are designed to be operated in a completely dry compartment (Fig. 237), while other models are equipped with brake assemblies that operate in oil (Fig. 240). The brakes on early models are actuated by mechanical linkage. Later models use two master cylinders and two slave cylinders to hydraulically actuate the brakes.

ADJUSTMENT

Models With Mechanical Brakes

190. To adjust mechanically actuated brakes, first be sure that both brake pull rods are adjusted to a length of 508 mm (20 inches). Turn each brake lever adjusting nut (Fig. 235) as required until brake pedal free play is 70 mm (2-3/4 inches) measured at pedal pad. Be sure both pedals are adjusted equally.

Road test and check operation of brakes. If there is a tendency for brakes to pull to one side, loosen adjusting nut

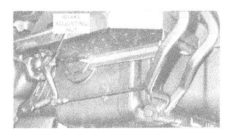

Fig. 235—Turn brake adjusting nut until pedal free play is 70 mm (2-3/4 inches) measured at pedal pad for mechanical brake models.

on side that pulls until brakes are equalized.

Models With Hydraulic Brakes

191. To adjust hydraulically actuated brakes, first make sure that brake pedals are at the top of their travel against the stops and that pedal latch can be freely latched between the pedals. Loosen locknuts on each master cylinder push rod (5—Fig. 241) and turn push rod into or out of clevis (6) until front end of push rod is just free to move. This ensures push rods are clear of master cylinder pistons and brakes are fully released.

Turn adjusting nut (1—Fig. 236) on each brake lever clockwise until brakes are fully engaged. Then, turn both adjuster nuts counterclockwise two turns. Make certain both rear wheels lock up at the same time with pedals latched together. If one wheel locks before the

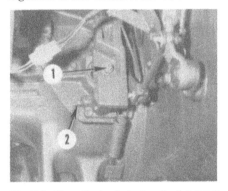

Fig. 236—View of hydraulic brake adjusting screw (1) and bleeder screw (2).

other, loosen the adjuster nut on ''locking'' brake until brakes apply evenly.

BLEEDING BRAKE SYSTEM

Models With Hydraulic Brakes

192. To bleed air from hydraulic brake system, first check level of brake fluid in reservoir and fill if necessary. Be sure to maintain full fluid level during bleeding procedure.

A bleeder screw (2—Fig. 236) is located on each brake slave cylinder. Attach a plastic tube to bleeder screw and place free end of tube into a container partially filled with brake fluid. Loosen the bleeder screw and fully depress pedal for brake being bled. Repeat this procedure until air-free fluid flows from submerged end of tube.

NOTE: If a plastic tube and container are not used as outlined above, bleeder screw must be closed before releasing brake pedal to prevent air from being drawn into system.

Be sure to tighten bleeder screw before removing plastic tube from bleeder screw or container. Repeat the procedure on slave cylinder on opposite side.

BRAKE DISCS AND ACTUATOR ASSEMBLY

Models With Dry Brakes

193. OVERHAUL. To remove brake assemblies, first remove complete final

Fig. 237—Exploded view of dry, disc type individual wheel brakes and control linkage used on some models.

1. Pedal shaft
2. Brake pedal
3. Parking pawl
4. Brake pedal
5. Brake rods
6. Brake lever
7. Lever support
8. Return spring
9. Woodruff key
10. Spacer
11. Cross shaft hinge bracket
12. Snap ring
14. Adjusting nut
15. Trunnion
17. Brake rod
18. Lined discs
19. Actuating disc

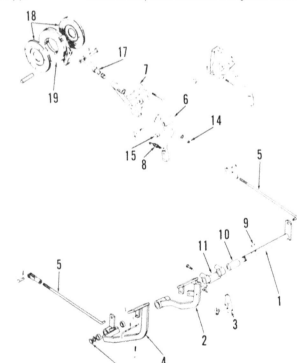

drive and axle housing units as outlined in paragraph 188. On right-hand side, remove differential lock fork and coupling (paragraph 181). Remove carrier plate from axle housing. Remove adjusting nut (14—Fig. 237) and trunnion (15) from end of brake rod (17). Unbolt and remove brake lever and support (7), then withdraw brake discs (18) and actuator (19) from axle housing.

Unhook the actuator return springs and separate actuator plates. Examine plates, actuating balls and ramps for wear or damage and renew as required. If any evidence of oil is present on brakes, find and correct the cause before reassembling brakes. Refer to paragraph 189 for axle housing oil seal renewal procedure.

To reinstall, reverse the removal procedure. Apply a light coat of a dry molybdenum disulphide lubricant to axle shaft splines where brake discs are located. Be sure there is no lubricant on any of the brake components.

Adjust brakes after assembly as outlined in paragraph 190.

Models With Wet Brakes

194. OVERHAUL. To remove brake assemblies, first remove complete final drive and axle housing units as outlined in paragraph 188. Remove brake adjusting nut (14—Fig. 240), then unbolt and remove brake lever (6) and lever support (7) as an assembly. Remove carrier plate and withdraw brake components from axle housing.

Disassemble actuator unit and inspect plates, balls and ramps for wear or damage and renew as required. Depth of grooves in friction plates (18) is 0.3 mm (0.012 inch) when new. Plates should be renewed if worn to the extent that grooves are beginning to disappear. Separator plates (20) should be renewed if excessively scored or worn. Separator

plates should also be renewed whenever friction plates are renewed. Renew plates that indicate signs of overheating or distortion.

IMPORTANT: Dip each friction plate, separator plate and the actuator unit in clean transmission oil before reassembling.

To reinstall brakes, reverse the removal procedure. Note that two friction plates and one separator plate must be installed on axle housing side of ac-

tuator unit (Fig. 239). Remaining friction plates and separator plates are to be installed alternately on carrier plate side of actuator. Install a new rubber boot (16—Fig. 240) making sure that garter spring end of boot seats properly in ridge of brake rod (17).

On models with mechanical brakes, adjust brakes as outlined in paragraph 190. On models with hydraulic brakes, bleed air from system as outlined in paragraph 192 and adjust brakes as outlined in paragraph 191.

Fig. 239—View of wet type disc brake components assembled onto carrier plate. Identification notches (N) indicate number of brake discs (2) used. Note that two brake discs and one separator plate are always installed on axle housing side of actuator assembly (1).

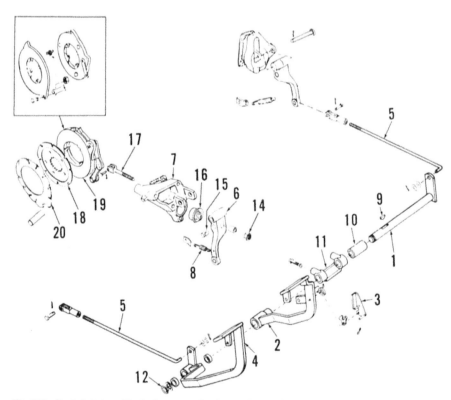

Fig. 240—Exploded view of typical wet type disc brake with mechanical actuating linkage used on some models. Brake actuator and brake discs are similar on models with hydraulically actuated brakes.

1. Pedal shaft	6. Brake lever	11. Cross shaft bracket	16. Boot
2. Brake pedal	7. Lever support	12. Snap ring	17. Brake rod
3. Parking pawl	8. Return spring	14. Adjusting nut	18. Friction disc
4. Brake pedal	9. Woodruff key	15. Trunnion	19. Actuator assy.
5. Brake rods	10. Spacer		20. Separator plate

Fig. 238—Carrier plate removed from left axle housing showing installed dry type brake assembly.

MASTER CYLINDERS

Models With Hydraulic Brakes

195. R&R AND OVERHAUL. To remove brake master cylinders (4—Fig. 241), first remove right-hand foot step. Disconnect supply pipes from master cylinders and drain reservoir or plug pipes to stop flow of brake fluid. Disconnect outlet pipes and plug ends of pipes. Disconnect clevis (6) from brake pedal bracket, remove mounting cap screws and withdraw master cylinder from the mounting bracket.

Remove rubber boot and push rod from master cylinder. Remove retaining ring (10—Fig. 242) and withdraw piston (9) and spindle assembly from master cylinder. Separate the spindle (3) from locking cap (7).

Master cylinder assembly should be renewed if cylinder bore or piston show signs of excessive wear, scoring or pitting. A seal kit is available for reassembly of master cylinder if piston and cylinder bore are in good condition.

To reassemble, reverse the disassembly procedure. After master cylinder is installed, bleed air from system as outlined in paragraph 192 and adjust brakes as outlined in paragraph 191.

SLAVE CYLINDERS

Models With Hydraulic Brakes

196. R&R AND OVERHAUL. To remove brake slave cylinders, first remove adjusting nut from end of brake actuating rod. Disconnect parking brake control rod. Disconnect return spring from brake actuating lever, then swivel the lever upward out of the way. Disconnect hydraulic pipe, remove retaining bolts and remove slave cylinder from actuator housing.

Remove rubber dust cover (7—Fig. 243) and plunger (6) from the cylinder. Remove retaining ring (5) and withdraw piston (3) from cylinder bore.

Renew slave cylinder assembly if cylinder bore or piston is excessively worn, scored or pitted.

Lubricate parts with clean brake fluid during assembly. Install piston seal (4) with flat side toward plunger end of piston.

Reinstall slave cylinders, bleed air from system as outlined in paragraph 192 and adjust brakes as outlined in paragraph 191.

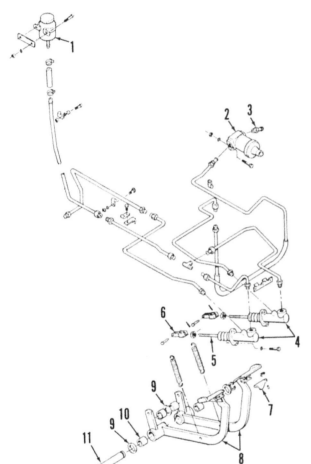

Fig. 241—Drawing of brake pedals, master cylinders, slave cylinder, reservoir and connecting pipes used on models equipped with hydraulic brakes. Late production models use dual reservoirs (1).

1. Reservoir
2. Slave cylinder (2)
3. Bleeder screw
4. Master cylinders
5. Push rod
6. Clevis
7. Latch
8. Brake pedals
9. Nylon washers
10. Bushing
11. Pedal shaft

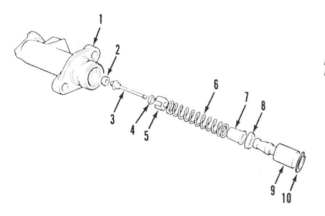

Fig. 242—Exploded view of hydraulic brake master cylinder.

1. Cylinder body
2. Spindle seal
3. Spindle
4. Spring washer
5. Seal cap
6. Spring
7. Locking cap
8. Piston seal
9. Piston
10. Retaining ring

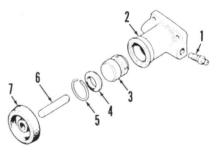

Fig. 243—Exploded view of hydraulic brake slave cylinder.

1. Bleeder screw
2. Cylinder body
3. Piston
4. Seal
5. Retaining ring
6. Plunger
7. Rubber boot

LIVE POWER TAKE-OFF

All models may be equipped with a live power take-off which may be operated at two speeds: proportional engine speed or proportional ground speed.

The pto control handle (1—Fig. 245) can be moved to engage the ground speed driven gear (7) with the drive gear on the bevel pinion shaft. This causes pto output shaft (10) to turn in direct proportion to ground speed of tractor.

The control handle can be moved to a midposition that disengages the pto drive.

A third (engine speed) position of control handle engages the pto output shaft with pto front drive shaft, which is driven from the dual stage clutch mounted on engine flywheel.

Depressing the clutch pedal about halfway stops motion of the tractor, but does not interrupt power flow to pto drive shaft. Completely depressing clutch pedal stops motion of the tractor as well as the pto shaft.

Refer to ENGINE CLUTCH section for service to dual clutch. Refer to appropriate TRANSMISSION section for service to pto input shaft and front drive shaft.

PTO OUTPUT SHAFT

All Models So Equipped

200. R&R AND OVERHAUL. The pto output shaft (10—Fig. 245) can be withdrawn from the rear as follows: Drain oil from differential housing. Remove shield (18) and pto cap (17). Unbolt and remove seal retainer plate (16). Pull the oil seal housing (14), output shaft and rear bearing (12) rearward from differential housing.

To renew needle bearing (9) at front of shaft, first remove left-hand axle housing and differential assembly. Use a slide hammer puller or other suitable tool to remove bearing from housing bore. Press against lettered side of new bearing using a suitable size stepped driver. Rear face of bearing should be just below chamfered edge of housing bore.

To renew ground speed gear (7) and/or bushing (8), it is necessary to remove the hydraulic lift cover and hydraulic pump. Press new bushing into housing bore until front edge of bushing is flush with inner end of housing bore chamfer. Be sure that shift lever (6) engages shift groove in ground speed gear when reinstalling gear.

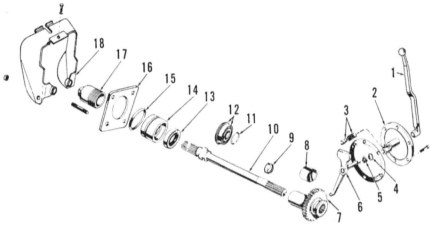

Fig. 245—Exploded view of live pto output shaft assembly and shift control unit used on some tractors.

1. Handle	9. Needle bearing	14. Retainer
2. Cover	6. Lever	10. Pto shaft	15. "O" ring
3. Detent	7. Ground speed driven	11. Snap ring	16. Retainer plate
4. Washer	gear/shift coupling	12. Bearing & snap ring	17. Cover
5. Seal	8. Bushing	13. Seal	18. Shield

INDEPENDENT POWER TAKE-OFF

All models may be equipped with an independent power take-off which is driven by a flywheel mounted "Split Torque" clutch and controlled by a hydraulically actuated, multiple disc clutch contained in the center of differential housing. A single speed (540 rpm) pto is available on all models, while a two speed (540 and 1000 rpm) pto is available in MF270 and MF290 models.

Refer to ENGINE CLUTCH section for service to Split Torque clutch. Refer to appropriate TRANSMISSION section for service to pto input shaft and front drive shaft.

OUTPUT SHAFT

Single Speed Models

203. R&R AND OVERHAUL. To remove ipto output shaft (10— Fig. 246), rear seal (13) and bearing (12), first drain

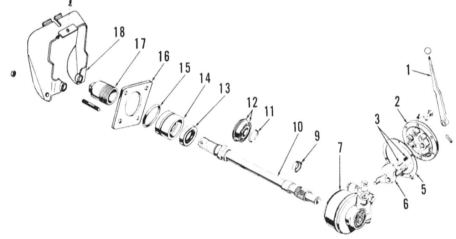

Fig. 246—Drawing of single speed independent pto output shaft (10), clutch (7) and associated parts used on some tractors.

1. Handle	6. Shift lever	11. Snap ring	15. "O" ring
2. Cover	7. Clutch assy.	12. Bearing & snap ring	16. Retainer plate
3. Detent	9. Needle bearing	13. Seal	17. Cover
5. Gasket	10. Pto shaft	14. Retainer	18. Shield

oil from differential housing. Remove shield (18) and retainer plate (16). Pull the seal housing (14) out with pto cover (17), then pull shaft and bearing rearward from differential housing.

Inspect all parts and renew if necessary. Be sure lip of oil seal faces forward. Front needle bearing (9) can be renewed after removing left-hand axle housing and differential assembly. Use a suitable piloted driver to install new bearing, pressing against lettered side of bearing. Press bearing into housing bore until rear edge of bearing is flush with rear face of housing bore.

To reinstall output shaft assembly, reverse the removal procedure.

Dual Speed Models

204. R&R AND OVERHAUL. To remove pto output shaft (15—Fig. 247), oil seal (17) and rear bearing (12), first drain oil from differential housing. Remove pto shield (23) and cap (21). Remove snap ring (16), then withdraw output shaft from the housing. Unbolt

and remove retaining plate (22). Pull seal retainer (20) with the oil seal and the sleeve (13) with rear bearing from housing bore. Remove seal (17) and metal shield (18) from retainer.

Refer to paragraph 205 for removal of pto driven gear (6), reduction gears (25) and needle bearings.

Press a new metal shield and oil seal into retainer housing making sure that lip of shield and lip of seal both face forward. Renew "O" rings (14 and 19).

To reinstall, reverse the removal procedure.

REDUCTION GEARS

Dual Speed Models

205. REMOVE AND REINSTALL. To remove dual speed pto reduction gears (25—Fig. 247) and pto driven gear (6), first drain oil from transmission housing. Remove hydraulic lift cover as outlined in paragraph 215. Split tractor between transmission housing and differential housing as outlined in

paragraph 156. Remove pto output shaft. Remove hydraulic pumps through front opening in differential housing and withdraw ipto clutch assembly (4). Remove snap ring (27), thrust washer (26), gear cluster (25) and bearings (24) from bevel pinion shaft. Remove pto driven gear (6) and thrust washer (7).

To renew needle bearings (8) it will be necessary to first remove left-hand axle housing and differential assembly. Drive the shaft guide tube (10) rearward from housing. Use a suitable puller to remove bearings and spacer (9) from housing bore. When installing new bearings, press against side of bearing stamped with part number. Press shaft guide tube (10) into housing until rear end of tube is no more that 1.5 mm (0.060 inch) past shoulder of counterbore for rear bearing (12).

To reinstall gears and related components, reverse the removal procedure.

IPTO CLUTCH AND CONTROL VALVE

All Models So Equipped

206. REMOVE AND REINSTALL. To remove clutch unit, first remove hydraulic lift cover as outlined in paragraph 215 and split tractor between transmission housing and differential housing as outlined in paragraph 156. Remove pto side cover. Disconnect hydraulic lines from auxiliary pump. Remove dowel pin from each side of differential housing retaining hydraulic lift pump, then slide pump assemblies forward and lift ipto clutch unit out of housing.

To reinstall clutch assembly, reverse the removal procedure.

Models With Early Style Clutch Unit

207. OVERHAUL. Refer to Fig. 248 for an exploded view of early style clutch unit. To disassemble clutch, push retainer plate (2) inward and remove retaining ring (1). Remove clutch discs (3), wave springs (4), separator plates (5) and drive hub (6). Remove piston (7) and seals (8 and 9) from clutch housing (11).

Unbolt and remove brake cylinder (20) and piston (19) from valve housing (14). Remove snap ring (16) and thrust washer (15), then withdraw valve housing. Brake disc (12) is retained to clutch housing by two pins (10).

To disassemble control valve spool, unseat internal snap ring (5—Fig. 249) from groove in plunger (2). Note that spacer ball (4) is available in alternate diameters and is used to establish compressed length of valve spool within recommended limits of 102.36-103.12

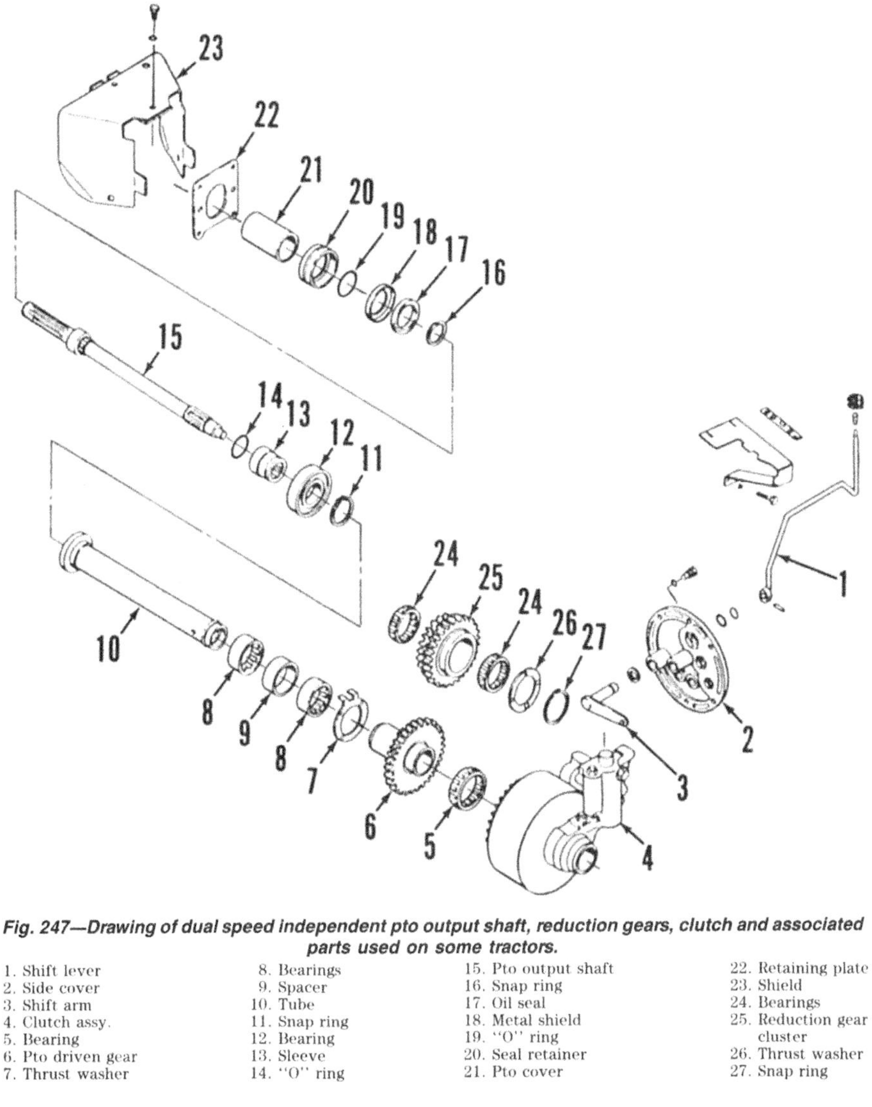

Fig. 247—Drawing of dual speed independent pto output shaft, reduction gears, clutch and associated parts used on some tractors.

1. Shift lever	8. Bearings	15. Pto output shaft	22. Retaining plate
2. Side cover	9. Spacer	16. Snap ring	23. Shield
3. Shift arm	10. Tube	17. Oil seal	24. Bearings
4. Clutch assy.	11. Snap ring	18. Metal shield	25. Reduction gear
5. Bearing	12. Bearing	19. "O" ring	cluster
6. Pto driven gear	13. Sleeve	20. Seal retainer	26. Thrust washer
7. Thrust washer	14. "O" ring	21. Pto cover	27. Snap ring

mm (4.030-4.060 inches) as shown in Fig. 250.

When assembling clutch unit, lubricate piston seal rings prior to installing piston in clutch housing. Several pieces of 5 mm (3/16 inch) diameter rod (T—Fig. 251) may be placed inside the housing as shown to serve as alignment guides for piston installation. Install drive hub (6—Fig. 248).

NOTE: Soak new clutch friction discs in clean transmission oil for 30 minutes before installing.

Beginning with an externally splined separator plate (5—Fig. 248), alternately install seven separator plates, six wave springs (4) and six friction discs (3). Push down on the last plate to compress the wave springs and insert two small rods into clutch housing bleed holes as shown in Fig. 252 to hold plates in place. Install remaining wave spring and friction disc, then install retainer plate (2—Fig. 248) and retaining ring (1). The restraining rods can be removed at this time.

Install new seal rings (13) on clutch hub, then assemble valve housing onto hub. Install thrust washer (15) and snap ring. Install valve spool in body.

Install brake disc (12) and roll pins (10). The pins should protrude 5 mm (3/16 inch) and be on opposite sides of clutch housing. Position wear plate (17) under brake disc, then attach brake cylinder to valve housing. Tighten mounting cap screws to a torque of 24 N·m (18 ft.-lbs.).

Models With Late Style Clutch Unit

208. OVERHAUL. Refer to Fig. 253 for an exploded view of late style ipto clutch and control valve assembly. To disassemble the removed clutch, first remove snap ring (20), thrust washer (19) and control valve housing (18) from hub of clutch housing (15). Remove retaining ring (4) and withdraw retainer plate (1), thrust washer (2), drive hub (3), shims (5), wave springs (6), separator plates (7) and friction plates (8) from clutch housing. Tap clutch housing against wooden work bench to remove piston front plate (10), Belleville washer (11), spacer (13) and clutch piston (14).

To disassemble control valve, remove detent plug, spring and ball (21). Drive out roll pin (23), then withdraw plunger (33) and valve spool (27) assembly from valve body. Unseat snap ring (32) and separate valve spool from plunger. Drive roll pin (29) from spring guide (30) and remove guide and spring (28) from valve spool. Remove retaining ring (24), then tap sleeve (26) with "O" ring from valve body bore.

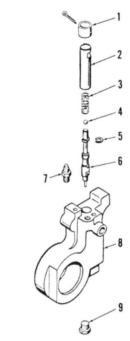

Fig. 249—Exploded view of early style ipto control valve assembly.

1. Retainer	
2. Plunger	6. Valve spool
3. Spring	7. Connector
4. Spacer ball	8. Housing
5. Retaining ring	9. Plug

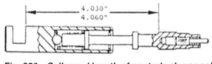

Fig. 250—Collapsed length of control valve spool should be 102.36-103.12 mm (4.030-4.060 inches) measured as shown. Length is adjusted by installing a different size spacer ball.

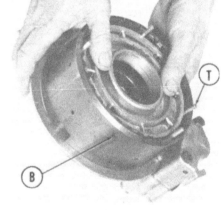

Fig. 251—Use pieces of 5 mm (3/16 inch) diameter rod (T) as alignment guides when installing piston in housing. Oil bleed holes (B) are used in assembly as outlined in text and shown in Fig. 252.

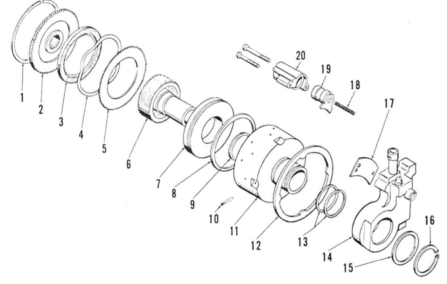

Fig. 248—Exploded view of ipto multiple disc clutch assembly used on early production tractors.

1. Retaining ring		
2. Retaining plate	7. Clutch piston	12. Brake disc	16. Snap ring
3. Friction disc	8. Piston seal ring	13. Seal rings	17. Wear plate
4. Wave spring	9. Piston seal ring	14. Control valve	18. Spring
5. Separator plate	10. Roll pin	housing	19. Brake piston
6. Drive hub	11. Clutch housing	15. Thrust washer	20. Brake cylinder

Fig. 252—Insert rods into bleed holes in clutch housing to hold separator springs compressed during clutch assembly. Refer to text.

Unbolt and remove brake pad (40) and brake cylinder (37) from control valve housing. Remove retaining ring (34) and withdraw piston (39) from cylinder.

Renew all "O" rings and seal rings. Renew clutch separator plates and friction plates if excessively worn, warped or discolored from overheating.

When assembling pto brake, a new shoulder bolt (41) should be installed to retain brake pad. If a new bolt is not available, apply Loctite 262 to threads of used bolt before installing. Tighten shoulder bolt and other mounting cap

screw to a torque of 6.8 N·m (60 in.-lbs.). Be sure brake pad moves freely.

To reassemble control valve, reverse the removal procedure. Install roll pin (23) with split side of pin facing away from plunger (33). Tighten detent plug to a torque of 47 N·m (35 ft.-lbs.).

When assembling clutch unit, lubricate piston seal rings with clean hydraulic oil. Install piston, flat side first, and the spacer ring into clutch housing. Install Belleville washer with concave (dished) side outward, and install piston front plate with stepped

edge inward. Beginning with an externally splined separator plate, alternately install seven separator plates, six wave springs and six friction discs. Push down on last separator plate to compress wave springs and insert two rods through bleed holes in clutch housing as shown in Fig. 252 to hold plates in place. Install remaining friction disc and wave spring, shims (if used), hub, thrust washer and retainer plate. Remove rods retaining clutch plates after snap ring is in place.

HYDRAULIC SYSTEM

The hydraulic lift system consists of a piston type pump which is submerged in hydraulic fluid in the differential housing and driven by the pto drive shaft. A control valve is located in the pump unit which meters the operating fluid at pump inlet. A single acting hydraulic cylinder, attached to bottom of lift cover, actuates the hitch rockshaft.

A gear type auxiliary pump supplies pressure for the Multi-Power clutch, ipto clutch and/or remote hydraulic cylinder applications on models so equipped. The auxiliary pump is mounted above the main hydraulic pump and is driven by a gear train located at the front of the pumps.

RESERVOIR AND FILTERS

All Models

210. The transmission/differential lubricant is the operating fluid for the hydraulic system. Manufacturer recommends renewing the hydraulic fluid after every 500 hours of operation, or annually, whichever comes first. Drain plugs (P—Fig. 255) are located on left-hand side of differential housing and transmission housing.

A reusable oil strainer element is located in bottom of differential housing at inlet of main hydraulic pump. Strainer element can be removed after first draining the oil and removing filter cover (6) from bottom of housing. The element should be cleaned using suitable solvent and compressed air after every 500 hours of operation. Some models are equipped with a warning light that comes on if strainer screen becomes plugged.

Some models are also equipped with an auxiliary hydraulic system oil filter element which is located in front of left-hand axle housing. The auxiliary filter element should be renewed after every 500 hours of operation.

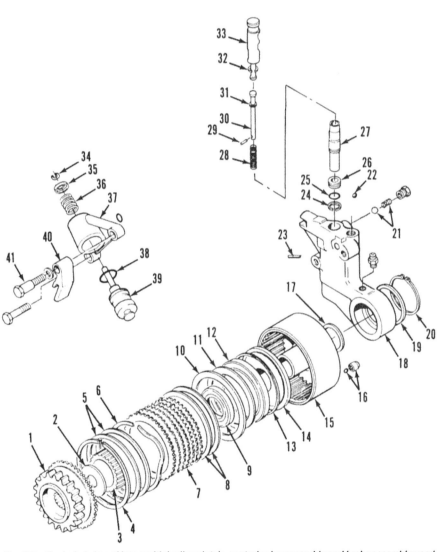

Fig. 253—Exploded view of ipto multiple disc clutch, control valve assembly and brake assembly used on late production tractors with two-speed ipto. Single speed pto unit is similar except gear is not used on rear of clutch retainer plate (1).

1. Retainer plate & gear	11. Belleville washer	21. Detent spring & ball
2. Thrust washer	12. Piston seal	22. Ball
3. Drive hub	13. Spacer	23. Roll pin
4. Retaining ring	14. Clutch piston	24. Retaining ring
5. Shims	15. Clutch housing	25. "O" ring
6. Wave springs	16. Check valve assy.	26. Sleeve
7. Separator plates	17. Seal ring	27. Valve spool
8. Friction discs	18. Control valve housing	28. Spring
9. Piston seal	19. Thrust washer	29. Pin
10. Piston front plate	20. Snap ring	30. Spring guide
		31. Washer

32. Retaining ring
33. Plunger
34. Retaining ring
35. Spring seat
36. Return spring
37. Brake cylinder
38. "O" ring
39. Piston
40. Brake pad
41. Shoulder bolt

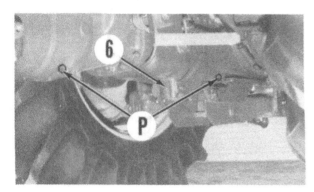

Fig. 255—Oil drain plugs (P) are located on left-hand side of transmission and differential housings. Oil strainer element is located above filter cover (6).

SYSTEM CHECKS AND ADJUSTMENTS

All Models

212. MASTER CONTROL SPRING. To check master control spring (7—Fig. 257) adjustment, first disconnect control beam at upper end and pivot the beam down out of the way. Check for end play by pulling and pushing on spring clevis (10). If end play is present, loosen Allen head set screw (3) in side of housing. Pull back rubber boot (9) to expose adjusting nut (8), then unscrew the nut and withdraw master control spring assembly from lift cover. Grasp spring (7) while holding clevis (10) and attempt to rotate spring on plunger (5). Spring should fit snugly with no end play but should still turn with moderate effort. If adjustment is incorrect, drive groove pin (11) out of clevis and thread clevis on or off plunger (5) until end play is just eliminated. Tighten plunger (5) if necessary until slot is aligned and reinstall pin (11).

Reinstall control spring assembly in lift cover. Move draft control lever to fully DOWN position, then tighten adjusting nut (8) until end play is just eliminated. Install a new nylon plug (2) and tighten set screw (3).

NOTE: End play will be present if adjusting nut (8) is either too loose or too tight.

213. INSTALLED ADJUSTMENTS. The following checks and adjustments should be performed with lift cover installed and tractor operational. Connect a 35000 kPa (5000 psi) pressure gage to test port (9—Fig. 258) in lift cover. Attach a weight of 400 kg (900 pounds) to the hitch lower links. Start the engine, then raise and lower the links several times to expel all air from the system. Perform system checks after oil has

TROUBLE-SHOOTING

All Models

211. The following are symptoms which may occur during operation of the hydraulic lift system and their possible causes. Use this information in conjunction with the System Checks and Adjustments information to isolate the cause of the problem before proceeding with component disassembly.

1. Hitch will not raise. Could be caused by:
   a. Internal oil leak in system. Remove response control side cover and check for leakage.
   b. Damaged, binding or misadjusted control valve linkage.
   c. Misadjusted or faulty servo valve (pressure control units), or faulty safety relief valve (nonpressure control units).
   d. Faulty hitch hydraulic pump.

2. Lift links raise unevenly (jerky). Could be caused by:
   a. Valve sticking in pump valve chamber.

3. Lift links will not raise to full height. Could be caused by:
   a. Transport stop misadjusted.
   b. Control valve misadjusted.
   c. Control linkage misadjusted.

4. Lift links will not lower. Could be caused by:
   a. Control valve sticking.
   b. Control valve misadjusted.
   c. Lift arms binding.

5. Hitch lowers too fast with response control at "SLOW" position. Could be caused by:
   a. Response control misadjusted.
   b. Control linkage binding or damaged.

6. Erratic action when operating in draft control. Could be caused by:
   a. End play in master control spring.
   b. Control linkage binding or damaged.

7. Lift links creep down. Could be caused by:
   a. Internal leakage in lift cylinder, control valve or valve chambers.

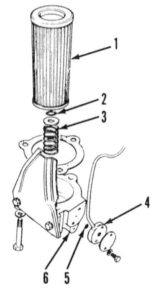

Fig. 256—Exploded view of oil strainer assembly used on models equipped with filter warning light. Note that narrow end of tapered element must be installed in cover (6).

1. Filter element
2. "O" ring
3. Spring
4. Warning light vacuum tube
5. "O" ring
6. Filter cover

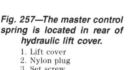

Fig. 257—The master control spring is located in rear of hydraulic lift cover.

1. Lift cover
2. Nylon plug
3. Set screw
4. Overload stop disc
5. Plunger
6. Spring seat
7. Control spring
8. Adjusting nut
9. Cover
10. Clevis
11. Groove pin

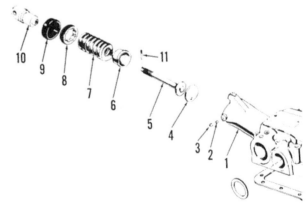

reached operating temperature of approximately 50° C (120° F).

To check system maximum pressure, set engine speed at 1500 rpm. Move draft control lever to fully UP position. For pressure control models, move position control lever to CONSTANT PUMPING position and observe pressure indicated by gage (7). For nonpressure control models, move position control lever past the transport stop (4) and observe pressure. On all models, maximum pressure should be 20340-21375 kPa (2950-3100 psi) and should not fluctuate more than 690 kPa (100 psi).

On nonpressure control models, maximum pressure is controlled by the safety relief valve (37—Fig. 295) located on lift pump housing. The relief valve is not adjustable.

On pressure control models, maximum pressure is controlled by the adjustable servo valve (4—Fig. 259) located on top of lift pump. To adjust pressure setting, remove rear cap screw from transfer plate and use a screwdriver through the opening to turn the servo valve adjusting screw as shown in Fig. 260. Each ½ turn will change pressure approximately 345-690 kPa (50-100 psi). If pressure fluctuates more than 690 kPa (100 psi), shut engine off, move position control lever fully down and remove response control side cover. Turn

diaphragm adjusting screw (1—Fig. 261) out ½ turn. Temporarily reinstall response control side cover and recheck for excessive fluctuation. Adjust as described ½ turn at a time until fluctuation stops, then turn screw out an additional one turn.

To check and adjust transport stop and position control setting, run engine at idle speed and move draft control lever to full UP position. Move position control lever to CONSTANT PUMPING

Fig. 260—On pressure control models, maximum system pressure is adjusted with a screwdriver through opening for rear cap screw of transfer plate.

section. When lift links are fully raised, scribe lines (L—Fig. 262) across lift cover casting and lift arm. Move position control lever to TRANSPORT position and measure the distance that scribed lines are separated. If separation is not within range of 1.6-4.7 mm (1/16-3/16 inch), loosen the two nuts (N—Fig. 263) securing inner quadrant and move inner quadrant and lever to obtain correct setting. Tighten the nuts and recheck setting.

To check and adjust draft control setting, start engine and set speed at 1200 rpm. Place position control lever in TRANSPORT position. Move draft control lever so lower links are suspended in a horizontal position and note location of control lever. If lever is not within sector marks (M—Fig. 263) on draft control quadrant, loosen screws (S) and relocate draft quadrant so lever is

Fig. 261—Diaphragm adjusting screw (1) can be adjusted after removing response control side cover. Oil should be drained to level of lower hole (6) before removing cover.

1. Diaphragm adjusting screw
2. Diaphragm lever
4. Response slide lever
5. Vertical lever

Fig. 258—View of controls for pressure control, 3-point lift system showing locations used in testing and adjustment. Gage (7) is used to check system pressure.

1. Draft control lever
2. Friction spring
3. Draft quadrant
4. Transport stop pin
5. Inner quadrant
6. Inner quadrant lever
7. Gage
8. Hose
9. Test port
10. Response control lever

Fig. 259—Lift pump controls for pressure control models as viewed through top opening with lift cover removed.

1. Diaphragm adjusting screw
2. Pump control valve roller
3. Response slide valve lever
4. Servo valve
5. Pressure adjusting screw & "butterfly" bracket

Fig. 262—Scribed lines (L) and sector marks (M) are used when adjusting lift controls (1 and 6). Refer to text.

within the sector marks. Be careful not to move position control quadrant. Tighten the screws and recheck setting.

To adjust response control, first remove outer cover from response control side cover. Loosen retaining screw (3—Fig. 264) and position response control lever 6 mm (1/4 inch) from rear end (slow position) of quadrant. Rotate inner lever (2) until internal cam contacts slide valve lever on the pump, then tighten retaining screw.

214. INTERNAL LINKAGE ADJUSTMENT. The following adjustment to internal lift linkage must be made before reinstalling lift cover. Two special tools (MFN970 gage and MFN1080D gage with Nuday No. 6088 and 6089 bushings) are required to properly make the internal linkage adjustments. On models equipped with flat seat deck and raised control quadrant, special tool (Nuday No. 12443) is also required. This tool is attached to outer end of quadrant support tube and is used to place position control and draft control linkage in proper position during adjustment.

Invert the lift cover in a fixture or on a bench, blocking up the cover so rockshaft can be moved to the normal raised TRANSPORT position. Make sure that master control spring is properly adjusted as outlined in paragraph 212. Place draft control lever in full UP position and place position control lever in TRANSPORT position. Adjust draft con-

trol rod stop screw (S—Fig. 265) to provide clearance (A) of 5.8 mm (0.230 inch) between head of screw and lift cover casting.

Install special tool MFN1080D (2—Fig. 266) on lift cover using special Nuday 6088 bushings (3) and two nuts and bolts. Be sure that the vertical lever (1) is to the rear of setting rod of special tool, then slide special Nuday 6089 bushing (B—Fig. 269) over the setting rod.

To adjust draft control linkage, first turn position control adjusting screw (5—Fig. 267) several turns out of position control cam linkage. Move position control lever to TRANSPORT position and place draft control lever in SECTOR position as shown in Fig. 268. Attach a suitable spring scale to vertical lever as

shown in Fig. 269 and apply a 1-1/2 kg (3 pounds) pull to vertical lever. The vertical lever should make light contact with bushing (B).

NOTE: Be sure that the straight edge of vertical lever (note the cutout) contacts the bushing.

If draft control setting is incorrect, straighten locking clip (6—Fig. 270) and loosen nut (7). While applying 1-1/2 kg (3 pounds) pull to vertical lever, move pivot bracket with a screwdriver until vertical lever makes light contact with bushing (B—Fig. 271). Tighten nut and lock it in place by bending locking clip when adjustment is correct.

To adjust position control linkage, move draft control lever to full UP position and move position control lever to TRANSPORT position. Install special Gage MFN970 (T—Fig. 272) on rear of lift cover with side of tool marked

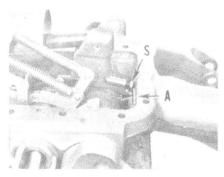

Fig. 265—Clearance (A) between draft rod cap screw (S) and lift cover casting should be 5.8 mm (0.230 inch). Refer to text.

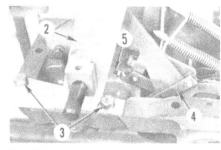

Fig. 267—Refer to text for internal adjustment to the lift cover linkage.

2. Tool MFN1080D
3. Bolts & Nuday 6088 bushings
4. Pressure control lever
5. Position control adjusting screw

Fig. 263—View of control quadrant for pressure control models.

M. Sector marks
N. Nuts
S. Screws
1. Draft control lever
2. Friction screw
4. Transport stop
6. Inner quadrant lever

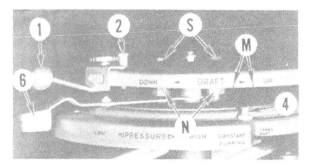

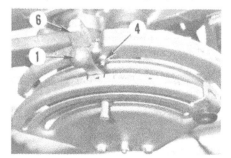

Fig. 268—View of quadrant for pressure control models. Refer to text.

Fig. 264—View of response control showing points of adjustment. Refer to text.

Fig. 266—The special tool (MFN1080D) shown at (2) should be installed using two Nuday 6088 bushings (3). Vertical lever is shown at (1).

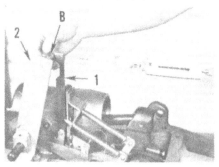

Fig. 269—Use a spring scale to apply 1-1/2 kg (3 pounds) pull to vertical lever (1) when adjusting draft control linkage. Refer to text.

''.636'' contacting rockshaft arm. Use spring scale to apply 1-1/2 kg (3 pounds) pull to vertical lever and check that vertical lever still makes light contact with bushing (B). If setting is incorrect, turn position control adjusting screw (5) as necessary until vertical lever contacts setting bushing. Tighten adjusting screw locknut, then recheck adjustment. Remove special tools after adjustments are completed.

On models equipped with raised control quadrant, be sure that draft and position control quadrant rods are adjusted to a length (L—Fig. 288) of 168 mm (6-5/8 inches), measured between the control rod upper and lower arms (3).

LIFT COVER

All Models

215. REMOVE AND REINSTALL. Disconnect lift links from lift arms and

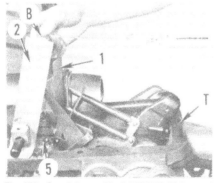

Fig. 270—Refer to text for changing position of pivot bracket (8).

Fig. 271—Use a screwdriver to move pivot bracket while checking for correct draft control setting.

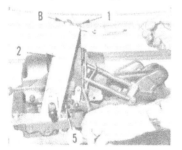

Fig. 272—Special gage block MFN970 (T) is used when setting position control linkage. Refer to text.

remove upper link control beam from master control spring clevis. Disconnect wiring, control linkage and hydraulic lines as necessary. Unbolt and remove seat, seat support deck and auxiliary hydraulic filter (if so equipped) as a unit as shown in Fig. 275. Drain oil from differential housing, then remove response control side cover (5—Fig. 276) and transfer plate (4) or selector control valve. Withdraw standpipe from lift pump. Remove attaching cap screws, install two guide studs, then remove lift cover from differential housing.

When installing lift cover, be sure that vertical control lever (4—Fig. 277) is to the rear of lever on hydraulic pump. Be sure that ''butterfly'' bracket (2—Fig. 278) is installed on top of pump servo valve (if so equipped) and carefully guide the pressure control lever (1) into ''butterfly'' as shown. A piece of welding rod can be used to guide standpipe into position as shown in Fig. 279. Tighten lift cover mounting cap screws to a torque of 88 N·m (65 ft.-lbs.)

Fig. 275—On models so equipped, auxiliary valve and return filter are removed with seat support as a unit.

Fig. 276—The response side cover (5) and transfer plate (4) or selector valve must be removed and standpipe withdrawn before removing hydraulic lift cover.

216. OVERHAUL. Refer to Figs. 280, 281 and 282 for exploded view of lift cover and linkage. Lift arms (23 and 28—Fig. 281), rockshaft arm (26) and rockshaft (27) all have master splines for correct assembly. Rockshaft bushings (25) are slip fit in lift cover bores.

To disassemble, loosen set screw (5—Fig. 280) and unscrew adjusting nut (13). Withdraw master control spring (12) assembly and disc (7) from lift cover.

Remove pressure control lever (50—Fig. 282) if so equipped. Loosen the position control adjusting screw (36—Fig. 283), then remove spring (38).

Fig. 277—The vertical lever (4) must be positioned as shown with lift cover installed.

1. Diaphragm adjusting
   screw
2. Diaphragm lever
3. Slide valve lever
4. Vertical lever

Fig. 278—When installing lift cover, be sure that pressure control lever (1) is positioned in ''butter-fly'' bracket (2).

Fig. 279—Standpipe (S) can be guided into position with a piece of welding rod as shown.

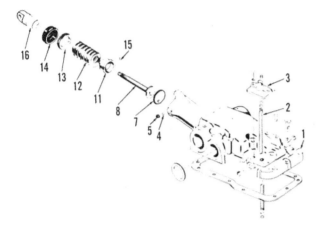

Fig. 280—Exploded view of master control spring typical of all models.

1. Lift cover
2. Standpipe
3. Transfer plate
4. Nylon plug
5. Set screw
7. Overload stop
8. Plunger
11. Spring seat
12. Control spring
13. Adjusting nut
14. Cover
15. Pin
16. Clevis

Fig. 284—Set screw (S) retains pivot shaft (P).

Thread a screw into quadrant retaining pin (P), then pull or pry pin from lift cover. Hold draft and position control fingers away from control cams as quadrant assembly is withdrawn from cover. Insert cotter pins through drilled holes in cam spring guide rods to hold springs (34—Fig. 282) in compressed position. Remove nut (29) retaining pivot bracket (31), then remove vertical lever (32) with fingers and pivot bracket as a unit. Remove plug from left-hand side of cover and loosen set screw (S—Fig. 284) retaining cam pivot shaft (P). Withdraw pivot shaft and remove position control and draft control cams (39 and 40—Fig. 282).

Remove right-hand lift arm (23—Fig. 281), then withdraw rockshaft (27) with left-hand lift arm from cover. Remove rockshaft arm (26) with piston rod (21). Withdraw piston (20) from cylinder (17) and unbolt and remove cylinder from cover if necessary.

When assembling, tighten lift cylinder retaining nuts to a torque of 160 N·m (120 ft.-lbs.). Grooved end of piston rod (21—Fig. 281) is retained in rockshaft arm by a set screw. If set screw was removed, reinstall using Loctite 262 on threads of screw. Turn set screw in until it bottoms, then back it out ¼ turn. Align master splines when installing rockshaft and lift arms. Tighten lift arm retaining cap screws until minimum shaft side play is obtained, but arms can still be raised and lowered freely.

Install control linkage as follows: Locate control cams (39 and 40—Fig. 282) in cover and start pivot shaft (P) through left-hand side of cover, but not into cams. Install vertical lever spring (41—Fig. 285) over pivot bracket (31) and around vertical lever (32), then insert pivot shaft into control cams. Do not tighten set screw at this time. Insert spring rods (34—Fig. 282) into mounting bracket (30), then remove cotter pins to release the springs. Install assembled quadrant while holding control fingers away from quadrant rollers. Install quadrant retaining pin (P—Fig. 283).

Hook spring (38) between position control cam and finger (37). Install pressure control lever (50—Fig. 282) if so equipped. Adjust cam pivot shaft (P—Fig.

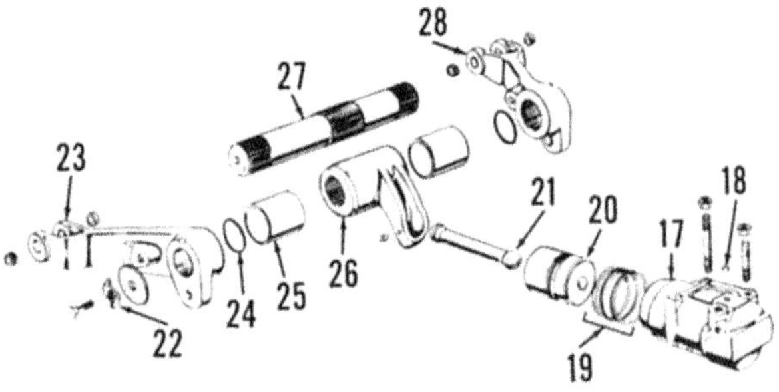

Fig. 281—Exploded view of hydraulic lift cylinder, rockshaft and associated parts.

17. Cylinder
18. "O" ring
19. Piston rings
20. Piston
21. Connecting rod
22. Lock clip
23. Lift arm
24. "O" ring
25. Bushings
26. Rockshaft arm
27. Rockshaft
28. Lift arm

Fig. 282—Drawing of hydraulic lift internal linkage typical of all models. Inset shows parts that are used for pressure control models.

P. Pivot shaft
S. Set screw
29. Nut
30. Support bracket
31. Pivot bracket
32. Vertical lever
33. Adjusting screw
34. Springs
35. Cam roller
36. Position control adjusting screw
37. Position control finger
38. Spring
39. Position control cam
40. Draft control cam
41. Spring
42. Draft control rod
43. Stop screw
50. Pressure control lever
51. "Butterfly" bracket

Fig. 283—Underside of hydraulic lift cover showing parts correctly installed.

P. Quadrant retainer pin
29. Nut
36. Position adjusting screw
37. Position control cam
38. Spring

284) to obtain minimum linkage side play, but make sure cams move freely. Tighten set screw (S) to secure pivot shaft.

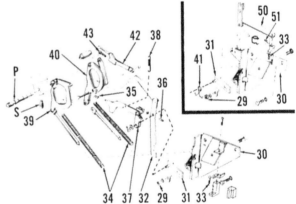

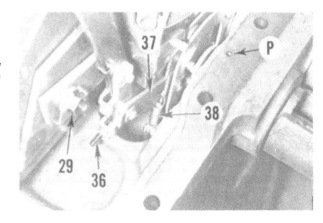

Install and adjust master control spring as outlined in paragraph 212. Adjust internal linkage as outlined in paragraph 214 before reinstalling lift cover.

CONTROL QUADRANT

All Models

217. On early models with auxiliary valve control levers located on left-hand side, the hitch control quadrant is mounted on outer end of quadrant support tube (3—Fig. 286). Draft and position control levers (9 and 10) are attached directly to the control shafts (1 and 2).

On late models with auxiliary valve control levers on right-hand side, the hitch control quadrant is mounted on

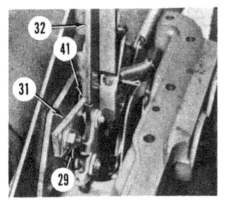

Fig. 285—View of spring (41) installed on vertical lever (32).

the seat deck. The draft and position control levers (9 and 10—Fig. 287) are connected to the control shafts (1 and 2) by control rods (19 and 21) and levers (20 and 22). Control rod length, measured between upper and lower arms (3—Fig. 288), should be set at 168 mm (6-5/8 inches).

The control quadrant support tube (3—Fig. 286 or Fig. 287) is retained in lift cover by a pin (P—Fig. 283) in bottom side of cover. With lift cover removed as outlined in paragraph 215, thread a screw into the retainer pin and pull or pry pin from cover. Remove pressure control lever (50—Fig. 282), if so equipped, and disconnect position control cam spring (38—Fig. 283). Hold draft and position control fingers away from quadrant cams, then rotate support tube back and forth while pulling it from lift cover.

To reinstall, reverse the removal procedure.

SELECTOR VALVE

Some models are equipped with a selector control valve which allows operation of lift linkage and auxiliary hydraulics simultaneously, or allows only the operation of auxiliary hydraulics.

With selector valve in "linkage" position, the lift linkage and auxiliary hydraulics may be operated independently. With selector valve in "auxiliary" position, the oil flow from hydraulic lift pump and auxiliary hydraulic pump is combined to provide maximum flow at the remote couplers. The lift linkage is locked and hitch cannot be used.

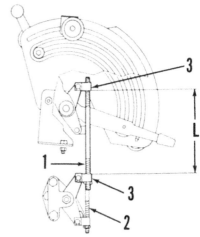

Fig. 288—Length (L) of raised control quadrant rods (1 and 2) should be set at 168 mm (6-5/8 inches), measured between upper and lower arms (3).

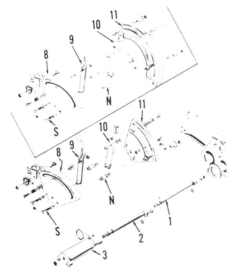

Fig. 286—Exploded view of typical control quadrant assembly used on all models except those equipped with raised quadrant. Inset shows parts for pressure control models. Refer to Fig. 287 for raised quadrant.

N. Special nuts
S. Screws
1. Draft control shaft
2. Position control shaft
3. Support
8. Draft control quadrant
9. Draft control lever
10. Position control lever
11. Position control quadrant

Fig. 287—Exploded view of raised control quadrant assembly used on late production models.

1. Draft control shaft
2. Position control shaft
3. Support
8. Draft control quadrant
9. Draft control lever
10. Position control lever
11. Position control quadrant
15. Transport stop
16. Lever friction adjusting screws
17. Retainer clip
18. Support brackets
19. Draft control rod
20. Draft lever
21. Position control rod
22. Position lever
23. Pivot pin

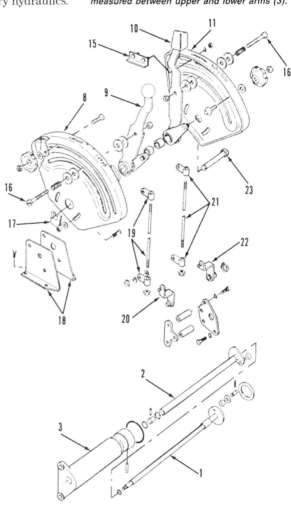

All Models So Equipped

220. LINKAGE ADJUSTMENT. To adjust selector valve control linkage, first move draft control and position control levers to DOWN position. Disconnect control linkage from selector valve bellcrank (7—Fig. 291). Rotate bellcrank clockwise until internal resistance is felt. Adjust lower control rod clevis (8) until retaining pin (9) fits freely into clevis and control valve bellcrank, then lengthen the rod by one turn of the clevis. Install the retaining pin and tighten clevis jam nut.

On models equipped with auxiliary control levers on left-hand side of tractor, adjust length of operating link (3) so force required to move hand control lever (1) to the external (auxiliary only) position is 11-18 N (2.5-4.0 pounds). Use a spring scale at outer end of control lever to measure force required to move lever.

On models equipped with auxiliary control levers on right-hand side of tractor, adjust the operating link screw (6) so force required to move hand control lever (4) to external (auxiliary only) position is 11-18 N (2.5-4.0 pounds). Measure force at end of handle using a spring scale.

221. R&R AND OVERHAUL. To remove selector valve, disconnect control linkage and hydraulic hoses from the valve. Remove the two mounting screws and lift valve assembly from lift cover.

To disassemble, unbolt and remove end plate (30—Fig. 291). Remove the detent (25), then pull bellcrank (7), sleeve (24) and actuator shaft (26) from valve body. Remove shift fork (29) and pull valve spool (28) from valve body.

Remove relief valve (23) assembly being careful to retain shims (22) for use in reassembly.

Clean and inspect all parts and renew if necessary. Valve spool (28) and body (27) are available only as an assembly. Relief valve (23) is used to protect lift

cylinder from shock loads when transporting heavy mounted implements over rough terrain with valve in external (auxiliary only) position. Relief valve is set to open at 26200-27600 kPa (3800-4000 psi). A suitable hydraulic test stand is required to check and adjust relief valve opening pressure.

To reassemble selector valve, reverse the disassembly procedure. Reinstall valve and tighten mounting screws evenly to 34-54 N·m (25-40 ft.-lbs.) torque. Adjust control linkage as outlined in paragraph 220.

HYDRAULIC LIFT PUMP

All Models

222. REMOVE AND REINSTALL. On models not equipped with ipto, Multi-Power or auxiliary hydraulics, the lift pump can be removed through top opening of differential housing after removing lift cover as outlined in paragraph 215. However, some

mechanics prefer to split the tractor between transmission and differential housing and remove pump out front opening of differential housing. With lift cover removed, remove "butterfly" bracket (1—Fig. 293) from upper tube (2) of pump servo valve (3) if so equipped. Collapse and remove the drive shaft and coupler connecting transmission output shaft to bevel pinion shaft. Remove pto output shaft. Remove pump mounting pins from both sides of differential housing, then lift pump out through top opening of housing.

On models equipped with ipto, Multi-Power or auxiliary hydraulics, tractor must be split between transmission and differential housings as outlined in paragraph 156 in order to remove lift pump. Remove lift cover as outlined in paragraph 215. Remove "butterfly" adjuster from upper tube of servo valve (3—Fig. 293) if so equipped. Remove pto side cover. Disconnect and remove hydraulic lines from auxiliary pump. Remove pump mounting dowels from

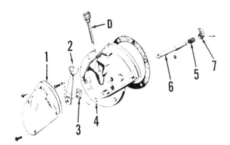

Fig. 290—Exploded view of response control and right side cover used on tractors not equipped with a selector valve.

D. Dipstick
1. Control cover
2. Response handle
3. Block

4. Side cover
5. Spring
6. Shaft
7. Response cam

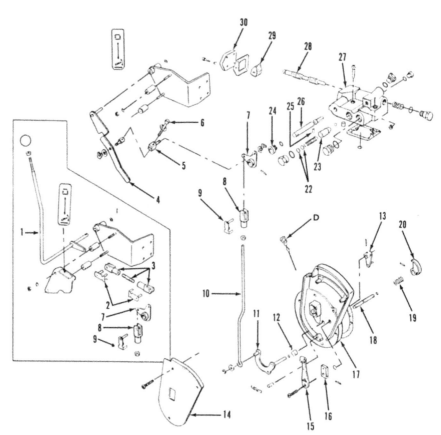

Fig. 291—Exploded view of response control, right side cover and selector valve assembly used on some models. Inset shows parts used on early models equipped with auxiliary valve control levers on left side of tractor.

D. Dipstick	7. Bellcrank	23. Safety relief valve
1. Selector control handle	8. Clevis	poppet
2. Retaining pins	9. Retaining pin	24. Sleeve
3. Control link	10. Lower control rod	25. Detent
4. Selector control handle	11. Lockout lever	26. Actuator shaft
5. Pivot	12. Bushing	27. Selector valve body
6. Adjusting screw	13. Arm	28. Valve spool
	14. Response control cover	29. Shift fork
	15. Response control handle	30. End plate
	16. Block	
	17. Side cover	
	18. Shaft	
	19. Spring	
	20. Response cam	
	22. Shims	

both sides of differential housing. Withdraw lift pump and auxiliary pump as a unit through front opening of housing as shown in Fig. 292.

To reinstall pump, reverse the removal procedure.

223. OVERHAUL. To disassemble the removed pump, first remove the two "C" clips attaching auxiliary pump drive train (if so equipped) to front of lift pump and separate auxiliary pump from lift pump. Remove the servo valve (30—Fig. 293), if so equipped, as follows: Remove upper tube (2) and spring from valve. Turn diaphragm adjusting screw (8) out until end of screw is flush with inside of pivot lever (7). Disconnect diaphragm lever link (6) from pump con-

Fig. 292—Hydraulic lift pump and auxiliary hydraulic pump are removed as a unit through front opening of differential housing.

trol lever (5). Remove nut from mounting stud, then use a brass drift to bump servo valve from pump housing.

Remove retainer clip (40—Fig. 294 or 295), then slide control lever (41) from control valve rollers. Remove rollers and pin from end of control valve (42). Remove filter element (36—Fig. 294) and cover (35), then unbolt and remove intake housing (34—Fig. 294 or 295) with cover plate (39) and slide valve assembly from pump rear body.

Remove end cap (1—Fig. 297) from front of oscillator drive, and remove pin (4) retaining oscillator (7) to cam follower (5). Remove nuts attaching pump bodies together, then slide front body (24—Fig. 294 or 295) off mounting studs. Withdraw control valve (42) and oscillator as a unit from rear body. Carefully slide piston yokes (28), valve chambers (25), cam blocks (26 and 30) and camshaft (31) from rear body (33). Separate cam blocks, piston yokes and valve chambers from camshaft.

Use a bronze welding rod with a hook formed on one end to pull valve sleeve (25—Fig. 296) from rear body. Remove "O" ring, back-up ring, spacer and washer from bore of rear body.

Overhaul the removed assemblies as outlined in the appropriate following paragraphs.

224. CONTROL VALVE. The control valve is serviced as a matched assembly which consists of rod (12—Fig. 297) and all parts (21 through 32).

To disassemble the valve and oscillator assembly, unseat and remove retaining ring (16) from internal groove in oscillator housing (7) and withdraw valve from oscillator unit. Remove retaining ring (8) while compressing spring (11). Push rod (12) in and remove snap ring (21), then withdraw remaining parts.

Assemble by reversing disassembly procedure. End of valve spool (24) with narrow slot should be toward oscillator rod (32); end with larger slots should be toward actuating rod (12).

Assemble spacer washer (17), spacer (18), back-up washer (19) and "O" ring (20) into rear body bore. Install sleeve (25) with "O" ring (27) and back-up washer (26) into body bore making sure that roll pin (23—Fig. 296) is located at bottom of bore. Outer surface of sleeve should be flush with machined surface of body when properly assembled.

225. VALVE CHAMBERS. Both valve chambers (25—Fig. 294 or 295) are identical and interchangeable, and each contains two sets of valve components (5 through 12—Fig. 298). To disassemble the chamber, remove snap ring (12) and thread a ¼ inch UNC bolt into plug (11) to provide leverage. Plug can be pulled by hand. A reseating reamer (Tool No. N-6007A) is available which dresses both seats in one bore at the same time. If tool is used, remove only enough metal to smooth the seats. All parts are available individually.

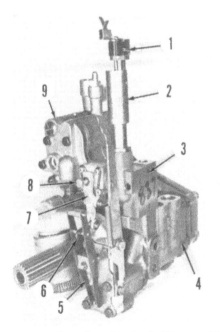

Fig. 293—Rear view of hydraulic lift pump (4) showing pump control linkage. Unit shown is equipped with pressure control servo valve (3) and auxiliary hydraulic pump (9).

1. Adjusting block assy.	6. Link
2. Tube	7. Diaphragm pivot
3. Servo valve assy.	lever
4. Lift pump	8. Diaphragm adjusting
5. Pump control valve	screw
lever	9. Auxiliary pump

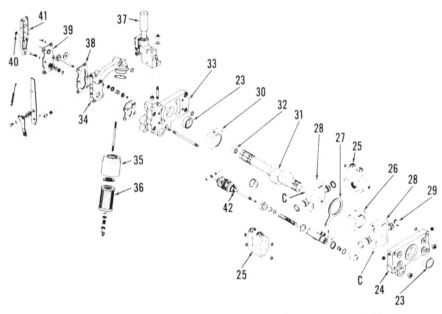

Fig. 294—Exploded view of hydraulic lift pump used on models with pressure control. Chamfered corners (C) of piston yokes face away from each other.

23. Bushing	28. Piston	33. Rear body	38. Gasket
24. Front cover	29. Piston rings	34. Intake housing	39. Rear cover
25. Valve chambers	30. Cam block	35. Filter bowl	40. Retainer clip
26. Cam block	31. Camshaft	36. Filter element	41. Actuating lever
27. Oscillator drive	32. Needle bearing	37. Servo valve	42. Control valve

When installing, do not compress retainer plug (11) deeper than 5.8 mm (0.23 inch) from top of plug to top of valve chamber. Be sure to pull plug up against snap ring (12) after snap ring is seated.

226. INTAKE HOUSING. The response control valve and the nonreturn valve are contained in the intake housing (34—Fig. 294 or 295). Refer to Fig. 299 for an exploded view of intake housing and components.

Be sure that "O" ring (9—Fig. 299), washer (8) and retaining clip (7) are correctly installed on cover plate (39). Install ring (11) on nonreturn valve (10), then insert valve through bore in housing. Install seal (1) over valve stem followed by retainer (2), spring (3), cap (4) and "E" ring (5). Install cover plate

(39) over gasket (38) and tighten the two attaching cap screws to 27 N·m (20 ft.-lbs.) torque.

227. SERVO VALVE. Refer to Fig. 300 for exploded view of servo valve. Unbolt and remove diaphragm cover (9). Use an Allen wrench to remove orifice plug (12). Thread a suitable screw into retainer plug (29) and remove retainer, spring (28), discs (27) and secondary relief valve (26). Remove guide (16) and spring support (17), then bump housing to remove plunger (20) and ball (20A). Use hooked wire to pull plunger guide (19) from bore then use needlenose pliers to pull ball valve seat (21) from bore.

Inspect all parts for excessive wear or damage and renew if necessary. Diameter of plunger (20), at large end,

should be 5.530-5.542 mm (0.2177-0.2182 inch). Inside diameter of plunger guide (19) should be 5.644-5.669 mm (0.2222-0.2232 inch). Renew ball valve seat (21) if pitted or scratched. Renew all "O" rings and back-up washers.

Install new back-up washer (22) and "O" ring (23) on ball valve seat (21), then install in bore of housing. Install new "O" ring (18) on plunger guide (19), then insert plunger guide into bore with cross drilled hole toward ball valve seat. Insert ball (20A) into bore and make sure that it enters bore of guide (19). Install plunger (20) with tapered (small) end toward ball. Install spring support (17) with large flat end against plunger guide (19). Apply 2 or 3 drops of "Loctite" grade 271 or equivalent sealant on threads of spring guide (16), then install and tighten to a torque of 40 N·m (30 ft.-lbs.).

Insert secondary relief valve (26), disc (27), spring (28) and retainer (29) into bore in bottom of housing. Press retainer in until flush with bottom of housing. Install orifice plug (12), diaphragm (11), spring (10), cover (9), lever (2) and related parts.

Adjust servo valve after pump is reinstalled as outlined in paragraph 213.

228. ASSEMBLY. When reassembling pump, be sure to renew all "O" rings and back-up washers and lubricate all parts with clean hydraulic oil. When installing new bushings (23) in pump bodies, be sure that split in bushing is located at the top and that chamfered surface of bushing does not extend above machined surface of bodies.

Assemble piston yokes into valve chambers using a suitable ring compressor to compress piston rings. Be sure that chamfered corners (C—Fig. 301) of piston yokes face away from each other. Install camshaft, cam follower and cam blocks into piston yokes making sure that cam block with shoulder is at front (widest lobe) of camshaft and that

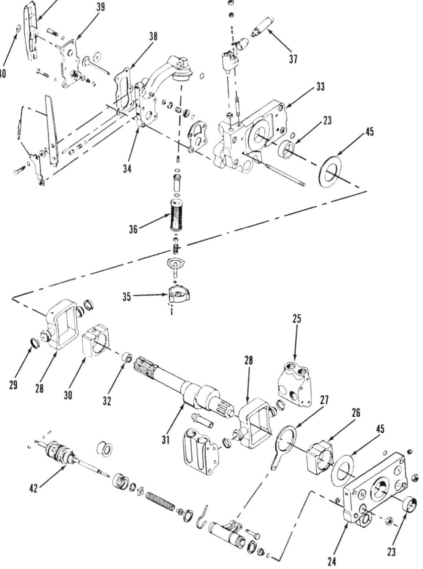

Fig. 295—Exploded view of hydraulic lift pump used on models without pressure control. Refer to Fig. 294 for legend except for the following: Preset safety relief valve (37) is used in place of servo valve. Thrust washers (45) are used on late production pumps.

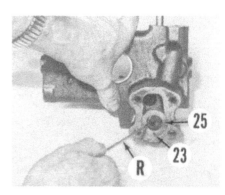

Fig. 296—Use a brass welding rod (R) with a hook on one end to pull spool sleeve (25) from rear body bore. Roll pin (23) must be located at the bottom when sleeve is reinstalled.

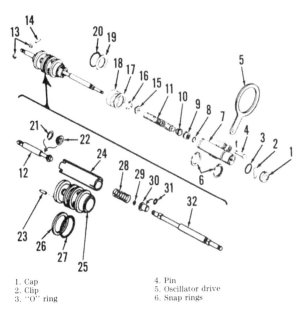

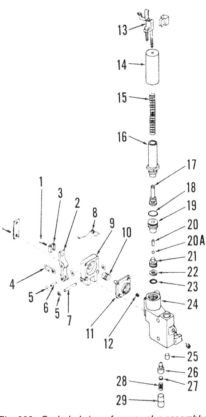

Fig. 297—Exploded view of hydraulic pump control valve and camshaft driven oscillator which prevents valve from sticking.

7. Oscilliator
8. Snap ring
9. Collar
10. Guide
11. Spring
12. Rod
13. Rollers
14. Pin
15. Disc
16. Snap ring
17. Washer
18. Spacer
19. Washer
20. "O" ring
21. Snap ring
22. Washer
23. Pin
24. Valve spool
25. Valve sleeve
26. Washer
27. "O" ring
28. Spring
29. Snap ring
30. Collar
31. "O" ring
32. Rod

1. Cap
2. Clip
3. "O" ring
4. Pin
5. Oscillator drive
6. Snap rings

shoulder faces inward. Slide camshaft assembly into rear body making sure that end of cam follower (27—Fig. 294 or 295) is toward bottom of pump.

Install control valve assembly (42) with oscillator into rear body, then position front body over camshaft and mounting studs. Install stud nuts making sure that nuts with grooves for auxiliary pump retaining clips (if so equipped) are installed on top right-hand stud (viewed from rear) and bottom left-hand stud. Tighten stud nuts evenly to a torque of 40 N·m (30 ft.-lbs.) while making sure that camshaft rotates freely. Install cap in bore of front body. Connect oscillator to cam follower.

Position intake housing onto rear body making sure that roll pin (23—Fig. 296) in valve sleeve engages slot in intake housing. Tighten the four retaining cap screws to a torque of 27 N·m (20 ft.-lbs.).

Assemble pin (14) and rollers (13) onto end of control valve with flat side of rollers against spool. Install pump control lever (41—Fig. 294 or 295) and retainer clip (40). Install servo valve assembly (if so equipped) and tighten mounting stud nut to a torque of 40 N·m (30 ft.-lbs.). Connect diaphragm pivot lever link to pump control lever. Turn diaphragm adjusting screw into pivot lever until three threads are visible on inner side of pivot lever to provide initial adjustment. Final adjustment of servo valve is made after pump is installed as outlined in paragraph 213.

Fig. 300—Exploded view of servo valve assembly used on pressure control models.

1. Adjusting screw	16. Guide
2. Lever	17. Support
3. Clip	18. "O" ring
4. Link	19. Sleeve
5. Clip	20. Plunger
6. Pin	20A. Ball
7. Pin	21. Seat
8. Spring	22. Washer
9. Diaphragm cover	23. "O" ring
10. Spring	24. Body
11. Diaphragm	25. Dowel
12. Orifice plug	26. Valve
13. Pressure adjusting bracket	27. Disc
14. Tube	28. Spring
15. Spring	29. Retainer

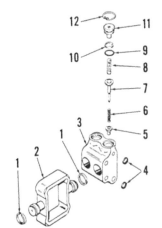

Fig. 298—Exploded view of valve chamber and associated parts.

1. Piston ring
2. Piston
3. Chamber
4. "O" rings
5. Inlet valve
6. Spring
7. Outlet valve
8. Spring
9. "O" ring
10. Back-up ring
11. Plug
12. Snap ring

Fig. 299—Exploded view of intake housing with nonreturn valve and response control slide valve components. Servo valve (SV) or safety relief valve is located on top of rear body (33).

1. Seal
2. Retainer
3. Spring
4. Cap
5. Snap ring
6. Gasket
7. Snap ring
8. Washer
9. "O" ring
10. Nonreturn valve
11. Ring
12. Pin
13. Orifice
14. Spring
15. Ball
16. Retainer
17. Bushing
18. Washer
19. Slide valve lever
20. Spring
21. "O" rings
22. "O" ring
23. Cap screws (short)
24. Cap screws (long)
33. Rear body
34. Intake housing
35. Filter housing
36. Filter element
37. "O" ring, washer & spring
38. Gasket
39. Cover
40. Retainer clip
41. Pump control lever

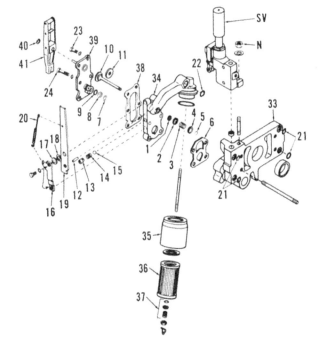

AUXILIARY HYDRAULIC PUMP

All Models So Equipped

230. Two different auxiliary pumps have been used. Single section pump shown in Fig. 303 is used on tractors equipped only with Multi-Power transmission and/or independent pto. Dual section pump shown in Fig. 304 or 305 is used on tractors equipped with auxiliary hydraulic system and also Multi-Power and/or independent pto.

On all models, the auxiliary pump is located directly above the hydraulic lift pump within the differential housing. The auxiliary pump is removed as a unit with lift pump through front opening of differential housing after splitting tractor between transmission and differential housings. Refer to paragraph 222 for removal procedure. Remove retaining clips (C—Fig. 302) and separate auxiliary pump and drive gear train from hydraulic lift pump. Disassemble gear train, being careful not to lose the 22 loose needle bearings (10) in the idler gear, and remove pump from mounting plate.

Drive gear train backlash is not adjustable; renew parts if backlash exceeds 0.40 mm (0.016 inch) between any two gears. When installing pump, tighten drive train retaining cap screws to a torque of 27 N·m (20 ft.-lbs.).

231. SINGLE PUMP. Refer to Fig. 303 for an exploded view of single element auxiliary hydraulic pump used on some models.

To disassemble pump, remove cap screws attaching end plate (1) to pump body (8) and separate pump components. Inspect all parts for wear, scoring or other damage and renew if necessary. All parts except pump body are available for renewal. Pump gears (5 and 6) must be renewed as a set. The gears will cut a wear track on inlet side of pump body bores during normal operation. If depth of track exceeds 0.10 mm (0.004 inch), pump assembly should be renewed.

Fig. 301—When assembling valve chambers onto piston yokes, be sure that chamfered corners (C) of yokes face away from each other.

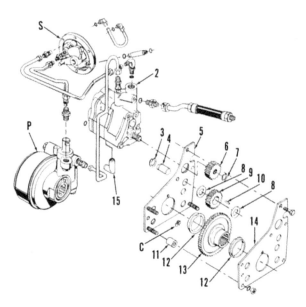

Fig. 302—Exploded view of auxiliary pump drive gear train. Unit shown is equipped with independent pto.

C. Retainer clip
P. Ipto clutch & valve assy.
S. Ipto control side cover
2. Auxiliary pump
3. Snap ring
4. Idler shaft
5. Mounting plate
6. Pump drive gear
7. Snap ring
8. Thrust washers
9. Idler gear
10. Needle roller
11. Spacer
12. Bushings
13. Drive gear
14. Mounting plate
15. Suction tube

The pressure plate (3) is hydraulically loaded to control gear end play. When installing pressure plate, be sure that relieved side of plate is on outlet side of pump. Tighten end cover retaining cap

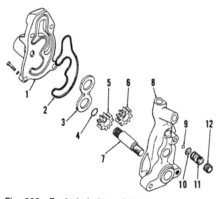

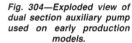

Fig. 303—Exploded view of single section auxiliary pump used on some models.

1. End cover	7. Drive shaft
2. Seal ring	8. Pump body
3. Pressure plate	9. Relief valve ball
4. Snap ring	10. Guide plate
5. Drive gear	11. Spring
6. Driven gear	12. Adjusting plug

Fig. 304—Exploded view of dual section auxiliary pump used on early production models.

1. Adjusting plug
3. Spring
4. Low pressure
  relief poppet
5. Ball
6. Front body
7. Back-up gasket
8. Seal
9. Wear plate
10. Tie bolts
11. Driven gear
12. Drive gear
13. "O" ring
15. Dowel
16. Rear body
17. Driven gear
18. Drive gear
19. Pin
20. High pressure
  relief poppet
21. Spacer plate
22. Spring
23. End cover
24. Adjusting screw

screws evenly to a torque of 27 N·m (20 ft.-lbs.).

The pressure relief valve (items 9 through 12) is set to open at 4825-6895 kPa (700-1000 psi). When removing valve for cleaning and inspection, count the number of turns required to remove adjusting plug (12) so the plug can be reinstalled to its original setting.

To check and adjust relief valve opening pressure, attach a hydraulic hand test pump with pressure gage to one of the pump outlet ports. Plug the other outlet port and restrain the pump gears from turning. Actuate hand pump and note relief valve opening pressure. Turn adjusting plug if necessary, to obtain specified opening pressure of 4825-6895 kPa (700-1000 psi). Stake the adjusting plug when correct pressure is obtained.

232. DUAL PUMP. Two different types of dual element auxiliary pumps have been used. Refer to Figs. 304 and 305 for exploded view of the pumps. Service procedures are similar for both pumps.

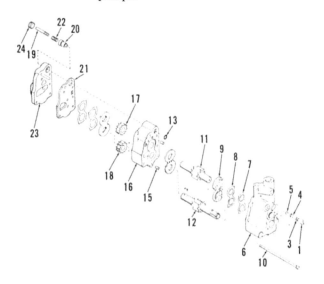

To disassemble the removed pump, first remove snap ring retaining drive gear (6—Fig. 302) to pump drive shaft. Remove cap screws attaching pump to gear train mounting plates and remove pump from rear plate (5). Remove tie bolts attaching pump bodies to end cover, then carefully separate pump bodies and gears.

NOTE: On late style pump, identify the bushings (27—Fig. 305) and the bores from which they were removed so they can be reinstalled in their original positions if reused.

When removing low pressure relief valve, count the number of turns required to remove the adjusting screw (1—Fig. 304 or 305) so it can be reinstalled to its original setting. On early style pump, do not remove high pressure relief valve adjusting screw (24—Fig. 304) from rear plate (23). On late style pump, high pressure relief valve (25—Fig. 305) is serviced as a unit and is not adjustable.

Check all parts for wear or damage and renew if necessary. If pump housings show evidence of wear or scoring in gear pocket area, complete pump assembly should be renewed. Pump gears must be renewed as pairs. Renew all seals, back-up gaskets and "O" rings. Lubricate all parts with hydraulic oil prior to assembly.

When assembling early style pump, be sure that pressure seal (8—Fig. 304) is installed into wear plate (9) first, then install back-up gasket (7). The seal side of wear plates must face away from gear faces. Omit high pressure relief valve spring (22) when assembling pump bodies. Tighten the tie bolts evenly to draw bodies together over dowel pins, then remove tie bolts and separate end cover (23) from spacer plate (21). Install high pressure spring (22) into bore in rear body, position end cover (23) on pump and tighten retaining bolts to 27 N·m (20 ft.-lbs.) torque.

When assembling late style pump, be sure that seal ring (8—Fig. 305) is installed on bushings (27) first, then install back-up washer (7). Install shaft bushings into body bores using hand pressure only. When installing front body bushings, be sure that upper bushing is positioned with relieved face on pressure port side of body. When properly assembled, face of rear bushings should protrude 0.0-0.5 mm (0.0-0.020 inch) above rear surface of front body (6), and face of rear gears (17 and 18) should protrude 0.0-0.5 mm (0.0-0.020 inch) above rear surface of rear body (16). Tighten the tie bolts evenly to a torque of 45 N·m (33 ft.-lbs.).

If low pressure relief valve was removed, reinstall and thread adjusting screw (1—Fig. 304 or 305) into front body the same number of turns required for removal. If original setting of adjusting screw is unknown, tighten adjusting screw until spring bottoms, then back screw out four turns. This should set relief valve opening pressure within specified range of 4825-6895 kPa (700-1000 psi).

Relief valve opening pressure can be checked as outlined in paragraphs 240 and 241 after pumps are reinstalled.

AUXILIARY VALVE

All Models So Equipped

235. Two different types of auxiliary control valves have been used. The valve assembly shown in Fig. 306 is used on early models and the valve shown in Fig. 307 is used on later models. Refer to the appropriate following paragraphs for service procedures covering each control valve.

236. EARLY PRODUCTION VALVE. Refer to Fig. 306 for an exploded view of 2-spool early style auxiliary valve. A single spool valve is also available. Service procedures are similar for either valve.

To disassemble, remove snap ring (16) and end plug (15). Loosen detent assembly (2) adjusting screws, then unbolt and remove end cap (1) from valve body. Withdraw spool assembly from spring end of valve body.

NOTE: Valve spools are selective fitted to their body bores and must not be interchanged. When servicing dual spool valve, identify spools as they are removed to ensure correct reassembly.

Remove retaining ring, centering spring (12) and spring seats from spool.

Remove screw (17), spring (18), poppet (19) and pin (22) or ball from regulator barrel (10). Unscrew regulator barrel from spool, then remove roll pin (20) and guide (23) from regulator barrel. Remove switch valve (6) and check valve (8) from valve body.

Inspect all parts for wear, scoring or other damage and renew if necessary. Valve spools and body are available only as a complete valve assembly. Be sure to renew all "O" rings.

To reassemble valve, reverse the disassembly procedure. Adjust spool detent and "kick out" pressure setting as outlined in the following paragraph.

237. ADJUSTMENT. A hydraulic flow meter must be used to properly adjust the auxiliary valve. Connect flow meter inlet hose to coupler No. 1 and connect flow meter outlet hose to coupler No. 2. Operate engine at 1000 rpm and move auxiliary valve control handle to "lift" position. Tighten detent adjusting screw (2—Fig. 306) until handle remains in "lift" position. Slowly close flow meter restrictor valve and note pressure at which valve spool "kicks out" of detent and control handle returns to neutral position. Spool should "kick out" of detent when pressure is between 13100-14480 kPa (1900-2100 psi).

If valve spool "kicks out" before pressure reaches 13100 kPa (1900 psi), shut off engine and operate control valve handle to relieve any trapped pressure. Remove snap ring (16) and end plug (15), then turn regulator valve adjusting screw (17) inward to increase "kick out" pressure.

NOTE: Several attempts at adjustment may be necessary before desired "kick out" pressure is obtained. Be sure to shut off engine and relieve system pressure before removing end plug from valve.

Fig. 305—Exploded view of dual section auxiliary pump used on late production models.

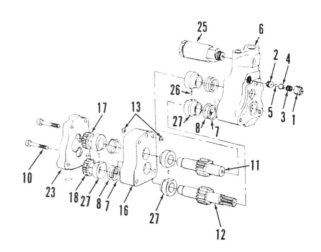

1. Adjusting plug
2. Ball seat
3. Spring
4. Low pressure relief poppet
5. Ball
6. Front body
7. Back-up washer
8. Seal ring
10. Tie bolts
11. Driven gear
12. Drive gear
13. "O" rings
16. Rear body
17. Driven gear
18. Drive gear
23. End cover
25. High pressure relief valve
26. Centering pin
27. Bushings

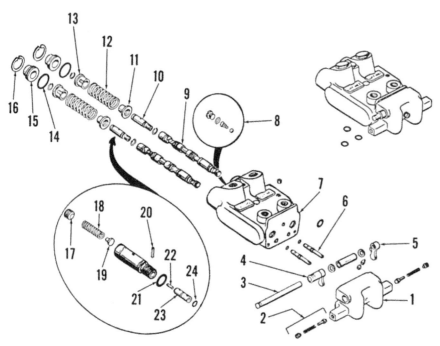

Fig. 306—Exploded view of 2-spool auxiliary valve used on early models. A single spool valve is also available and is similar in design.

1. End cap	7. Body	13. Sleeve	19. Guide
2. Detent assy.	8. Check valve	14. "O" ring	20. Pin
3. Inner rod	9. Spool	15. Plug	21. "O" ring
4. Outer lever	10. Spool plug	16. Snap ring	22. Pin
5. Inner lever	11. Sleeve	17. Nylon screw	23. Guide
6. Switch valve	12. Spring	18. Spring	24. "O" ring

239. ADJUSTMENT. A hydraulic flow meter must be used to properly set valve spool detent "kick out" pressure. Connect flow meter inlet hose to remote outlet No. 1 and connect flow meter outlet hose to outlet No. 2. Operate engine at 1000 rpm and move control valve lever rearward to "lift" position. Slowly close flow meter restrictor valve and note pressure at which control valve "kicks out" of detent and handle returns to neutral. Valve should "kick out" of detent at approximately 16550 kPa (2400 psi). "Kick out" pressure is adjusted by turning detent relief valve adjusting screw (20—Fig. 307) in to increase pressure or out to decrease pressure.

HYDRAULIC TESTING

All Models So Equipped

240. AUXILIARY PUMP HIGH PRESSURE. The auxiliary hydraulic system maximum pressure can be checked at an auxiliary valve remote outlet using a 35000 kPa (5000 psi) pressure gage and suitable fittings. Start engine and operate until hydraulic oil temperature is at least 50° C (120° F). Actuate auxiliary valve control lever and observe pressure gage reading.

If valve spool does not "kick out" of detent when pressure reaches 14480 kPa (2100 psi) and a "squealing" noise from control valve is heard, loosen control handle detent screw (2) until handle releases within specified pressure range. If valve does not "kick out" of detent and no "squealing" sound is heard, shut off engine and remove end plug (15) from valve. Turn regulator adjusting screw (17) outward to decrease "kick out" pressure. Reinstall plug and recheck for proper operation.

238. LATE PRODUCTION VALVE. Refer to Fig. 307 for an exploded view of single spool auxiliary valve used on some models.

To disassemble auxiliary valve, remove single/double acting selector valve (25—Fig. 307), check valve (23) and detent pressure valve assembly (21) from valve body (24). Remove end cap (19), unscrew retaining screw (18) and remove centering spring (16), spring seats (15) and end plate (14). Remove pivot bracket (4) with wiper seal (5) and "O" ring (6). Pull spool (3) out of valve body and remove detent carrier (11), piston (8) and cage (7).

Inspect all parts for wear and renew as necessary. Valve spool and body are available only as a matched assembly. Renew all "O" rings when reassembling.

When reassembling, use petroleum jelly to hold detent balls in place. Clean threads of retaining screw (18) and valve

spool, then apply a drop of Loctite 271 to threads of screw before installing.

Check and adjust valve spool detent "kick out" pressure as outlined in paragraph 239.

Fig. 307—Exploded view of single spool auxiliary valve assembly, inlet cover and outlet cover used on late models. A second single spool valve unit can be installed in the "valve stack" to provide dual auxiliary valves.

1. Inlet cover
2. Pivot link
3. Valve spool
4. Pivot bracket
5. Wiper seal
6. "O" ring
7. Detent cage
8. Piston
9. Spring
10. Detent balls
11. Detent carrier
12. "O" ring
13. "O" ring
14. End plate
15. Spring seats
16. Centering spring
17. Spacer
18. Shoulder bolt
19. End cap
20. Detent adjusting screw
21. Detent valve assy.
22. Outlet cover
23. Check valve assy.
24. Valve body
25. Single/double acting selector valve

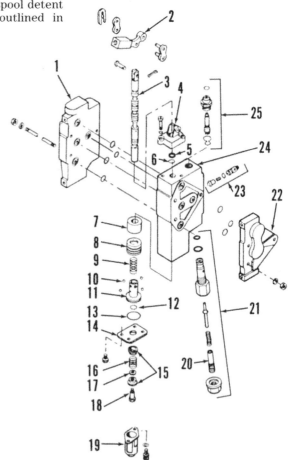

Relief valve opening pressure should be 15860-17925 kPa (2300-2600 psi) on Models MF255, MF265 and MF275. On Models MF270 and MF290, relief pressure should be 17240-19305 kPa (2500-2800 psi). Hydraulic lift cover must be removed as outlined in paragraph 215 for access to auxiliary pump high pressure relief valve.

All Models So Equipped

241. AUXILIARY PUMP LOW PRESSURE. To check auxiliary pump low pressure relief valve setting, disconnect pressure line to oil cooler (located at top left-hand side of transmission housing). Install a 10000 kPa (2000 psi) pressure gage in pressure line and cap end of line going to oil cooler. Place ipto lever in disengaged position, start engine and observe pressure gage reading. Pressure should be 4825-6895 kPa (700-1000 psi). Hydraulic lift cover must be removed as outlined in paragraph 215 for access to auxiliary pump low pressure relief valve.

All Models So Equipped

242. MULTI-POWER PRESSURE TEST. To check Multi-Power system pressure, disconnect oil cooler supply and return hydraulic lines at top left-hand side of transmission housing. Connect inlet hose of a suitable hydraulic flow meter to supply line (pipe nearest left-hand side of tractor). Connect outlet hose of flow meter to return pipe. Open flow meter restrictor valve and operate engine at 2000 rpm. With Multi-Power in HIGH or LOW, pressure should be 1585-1860 kPa (230-270 psi).

Multi-Power pressure regulator valve is contained in the Multi-Power shift valve which is mounted on transmission input shaft retainer housing. Refer to paragraph 170 to service the control valve assembly.

All Models So Equipped

243. IPTO PRESSURE TEST. To check ipto operating pressure, install a 4000 kPa (600 psi) pressure gage in test port in pto side cover on left-hand side of differential housing (Fig. 308). Start engine and run at 2000 rpm. Oil temperature should be 50°-60° C (120°-140° F).

Move pto control lever rearward to engaged position and note pressure gage reading. On models with single speed pto, pressure should be 1895-2240 kPa (275-325 psi). On models with two speed pto, pressure should be 1550-1895 kPa (225-275 psi). Move control lever forward to disengaged position. Pressure should drop to zero and pto shaft should stop in about five seconds.

All Models So Equipped

244. PRESSURE MAINTAINING CHECK VALVE. Late production tractors equipped with eight speed transmission and independent pto use a pressure maintaining valve located in pressure line to the oil cooler, to maintain back pressure in hydraulic system. The back pressure ensures that oil under pressure is available to operate independent pto clutch pack.

To check pressure maintaining valve operating pressure, install a 4000 kPa (600 psi) hydraulic test gage into oil cooler pressure line using a tee fitting. Operate engine at 1000 rpm and note pressure gage reading. The pressure at which valve regulates is stamped on the valve body. Pressure maintaining valve can be removed after removing front grille and hood and disconnecting hydraulic lines from the valve. Unscrew adapter fitting (6—Fig. 309) and remove sleeve (4), spring (3) and poppet (2). Inspect poppet and seat and renew if necessary.

Be sure to note direction of oil flow when reconnecting valve to oil cooler lines.

Fig. 308—Install a 4000 kPa (600 psi) pressure gage in test port in left-hand side cover to check ipto clutch pressure.

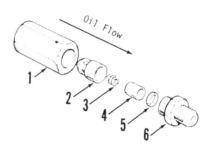

Fig. 309—Exploded view of pressure maintaining valve used on some models equipped with independent pto and eight speed transmission.

1. Valve body	4. Sleeve
2. Poppet	5. Ring
3. Spring	6. Adapter fitting

WIRING DIAGRAMS

All Models

245. Refer to appropriate Fig. 310, 311 or 312 for tractor wiring diagram. Service procedures covering electrical system components are contained in paragraphs 135 through 146.

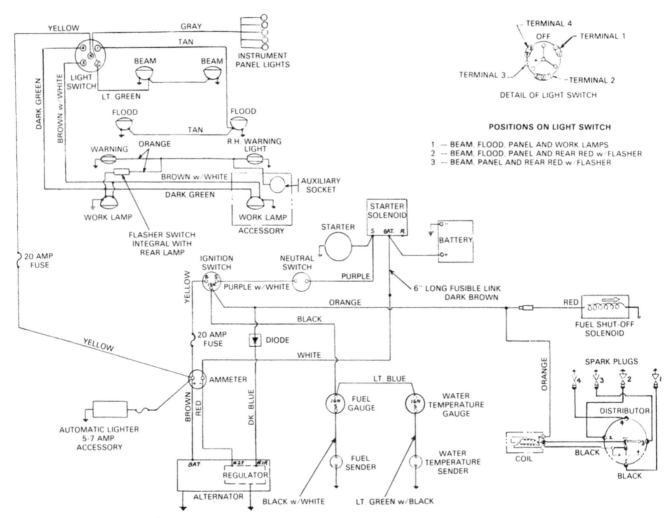

Fig. 310—Wiring diagram typical of Models MF255 and MF265 with gasoline engines.

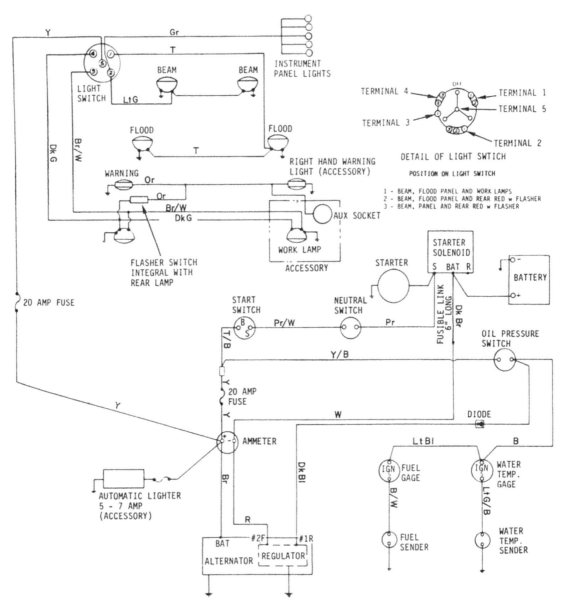

Fig. 311—Wiring diagram typical of Models MF255, MF265 and MF275 with diesel engines.

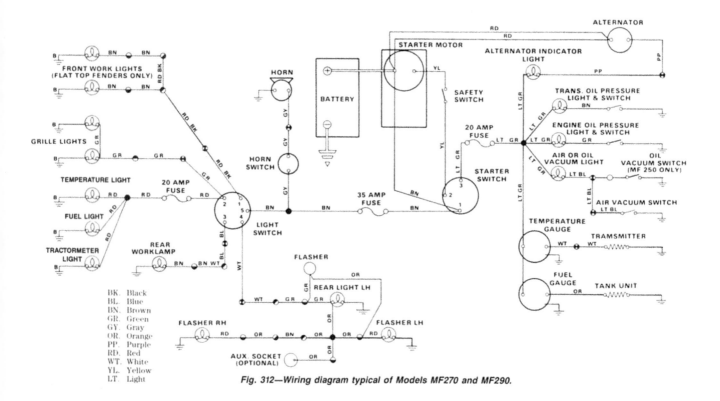

BK. Black
BL. Blue
BN. Brown
GR. Green
GY. Gray
OR. Orange
PP. Purple
RD. Red
WT. White
YL. Yellow
LT. Light

Fig. 312—Wiring diagram typical of Models MF270 and MF290.

NOTES

NOTES